ABBEY PACHTER

A
MONARCH
IN WINTER

BIOGRAPHY OF A
BUTTERFLY

Publishing Services provided by Paper Raven Books LLC
Published by Abbey Pachter
Printed in the United States of America
First Printing, 2023

Cover design: Lizaa
Book design: Heather Preis and Lynessa Layne
Author photo: Madeline Szul

Paperback ISBN 979-8-9888676-0-9
Hardback ISBN 979-8-9888676-1-6
eBook ISBN 979-8-9888676-2-3

To my grandkids Nico, Travis, Rhys & Anna:
Keep looking. Nature is a wonderful teacher.

Contents

Part 1

January

For admirers of something barely born, (not to help) goes against the sacred nature of our calling.

—*Call the Midwife* Season 2 Episode 2[1]

New Year's Day, 2021

Maybe it's dead.

I padded downstairs shortly after six, cozy in shearling slippers and a still-fluffy pink bathrobe that was a gift from my daughters so long ago that I'd had its sleeves cut and re-hemmed above their frayed cuffs. While almond milk warmed in the microwave for my morning cocoa, I shuffled from the kitchen to the still-dark living room to check a small table by the corner windows where I'd left several stalks of milkweed leaves, a twig, and a monarch chrysalis hanging from it, all in a tall, glass vase. At first glance, I was startled to see that overnight the chrysalis had turned black. Maybe I should have left it outdoors where it would have quickly frozen to death.

More than two weeks ago, I found the season's last monarch caterpillar in my garden. Its two companions, seen the day before, had disappeared to hide, pupate to rest for the winter, or die, frozen. That morning's rain had turned to sleet. The green, yellow, white, and black, circumferentially striped caterpillar clung to the milkweed, motionless from the cold, encased in a shiny layer of ice. Compelled by deep instinct to rescue it, I brought it indoors along with some milkweed stalks to eat if it survived. I smiled with the pleasure of success when, an hour later it had warmed up enough to inch along the stalks and munch on the milkweed leaves. After several days of feeding on them, the caterpillar became a chrysalis. When the new year arrived at midnight, it resembled a jewel-like green pendant with a collar of tiny gold pearls and scattered speckles of golden dots.

Chrysalis

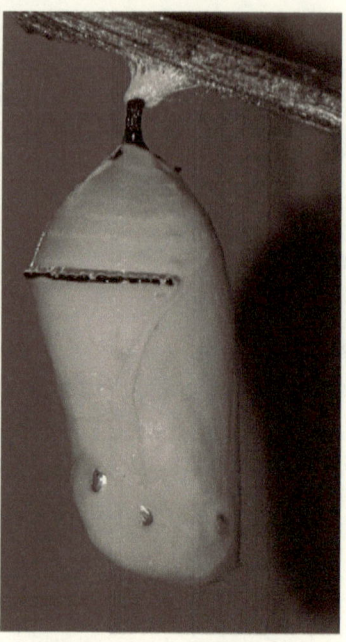

One year after COVID-19 arrived in the USA, it was still sickening hundreds of people daily. To decrease the risk of contagion, friends rarely met in person. We used the computer application Zoom, unfamiliar before the pandemic, to reduce the strain of quarantine. I hadn't seen one daughter and her family for ten months. Weekly Zoom visits with both daughters and the rest of our family helped us stay connected. I ached from missing them but was reassured by the isolation, mask-wearing, and hand-washing protocols they followed to avoid contracting the deadly virus. Although a new vaccine was recently approved, its administration hadn't begun.

The holiday season passed quietly. Chanukah brought bits of light against winter's darkness. The chrysalis held a note of renewal. Its gold spots sparkled in the candlelight.

Checking the chrysalis, a daily occupation, distracted me from the still-frightening COVID-19 statistics and persistent intransigence of deniers. Along with at least half the people in the USA and millions more worldwide, I waited hopefully for the new vaccine to become available.

The new year held promise. A new president would be inaugurated soon. The awful year, 2020, passed into history.

Happy New Year

When I got closer to the table with its vase and chrysalis, I saw something promising. Although at a distance it looked all black, when I leaned in for a closer look, I saw that although it was dark, it was nearly transparent. Still, this was a marked change from its previous green opacity. I wasn't sure what had happened. Bits that resembled wings showed through. Wavy black stripes were interlaced with irregularly shaped streaks of amber and scattered spots of white. After a short think, my mind dredged up the memory—these were signs that its transformational metamorphosis was almost complete. Years before, while studying to become a Master Naturalist, I'd learned the proper name for the emergence of a butterfly from its chrysalis.

5

Maybe it will eclose soon.

In the remarkable way memories' synapses connect disparate events, my thoughts jumped back in time. Six years ago, my daughter was pregnant, overdue, and increasingly uncomfortable carrying their first child—my first grandbaby. After forty years of experience as a women's healthcare professional, I'd witnessed many a pregnant belly. As a former Certified Nurse-Midwife, I'd helped nearly three hundred women deliver. I could practically see our little one squirming and kicking from being all flexed so tightly in her beautifully large belly, waiting to answer life's mysterious call to birth.

What happened?

Week 1

January 2
Morning

It's here! Autumn's last caterpillar has become the first butterfly of the new year.

When I made my morning coffee, it was still a chrysalis. An hour later, when I came back downstairs after having showered and dressed, it had become a butterfly.

Although I'd missed the moment, I felt a smile wrinkle my cheeks upon seeing the result. It was motionless, hanging with its wings dangling. Gravity had already pulled plasma-like hemolymph through veins and cells to expand its wings into their characteristic shape. They looked as velvety as a newborn's skin freshly washed from the cheesy vernix that protected it from the amniotic waters of pregnancy.

While it rested motionless, those newly expanded wings dried and became firm and sturdy. I left it alone since any interruption during the first several hours after its arrival could abort the successful completion of these changes. A fall caused by wind, the bite of a predator, or a human grabbing it could result in a useless wing, which would leave the butterfly unable to fly, feed itself, or mate. Its only purpose, in that case, would be as food—or mulch.

I marveled at what had successfully occurred since it had become a chrysalis. During the two weeks of its metamorphosis, everything that had been a caterpillar dissolved into a pool of stem cells, hormones, and DNA that rearranged itself to create a butterfly. Entomologists have not yet completely demystified how this happens. It's still a discoverable miracle of nature.

Although I'd watched YouTube movies of monarchs eclosing with their still-crumpled wings vulnerably soft and flightless, I was sorry to have missed seeing it happen in real life. I consoled myself by reflecting on the rarity of seeing a butterfly emerge in the wild. If less hidden, such a new, moist butterfly could make a tasty morsel for a watchful spider, lizard, or mouse.

Recently eclosed

I'd attended hundreds of human pregnancies and births when I was a Certified Nurse-Midwife and taught these subjects to hundreds of nursing students. While I gazed pensively at this new butterfly, I could almost feel my brain buzzing with electrical impulses that jumped across synapses between memories and suggested evolutionary parallels.

Lungs and wings.

In a pregnant woman's uterus—like in a chrysalis—space is at a premium. When the developing fetus and butterfly approach full term, parts not yet needed are squashed, including fetal lungs and butterfly wings until both emerge from their protective but constricted surroundings.

When no longer confined, those young parts expand. Decompressed, muscles in the chest assist breathing. Alveoli, the tiny sacs within human lungs, expand with a newborn's initial breaths, inflating a little more with each one taken during the first moments after birth. Most often, they can perform their essential functions of exchanging oxygen and carbon dioxide.

Breathing creates pressure changes in the lungs, heart, and blood vessels that render obsolete the placenta, umbilical cord, and fetal circulatory structures, including the foramen ovale, ductus arteriosus, and ductus venosus. The cord is clamped and cut, the placenta leaves the uterus, and the baby is freed, like a spaceship separating from spent launch rockets.

I realized—with surprise—that butterflies undergo related processes. I wanted to learn more, and am indebted to a local entomologist who advised me to look in Chapman's[2] insect physiology book. It proved enormously helpful and readable. I read in that when butterflies emerge, tiny muscles in their spiracles—small, trachea-like tubules spaced along the abdomen—expand and contract. Freed from the confines of the chrysalis, they take in oxygen and expel carbon dioxide.

When a butterfly climbs out of its chrysalis, it must grasp with its feet and hang with its wings down so gravity assists the wing's expansion. Once the paper-thin wings are fully opened, the enabling venous pathways desiccate. The wings harden into their strong, flight-enabled state.

Babies breathe; butterflies fly.

I pulled an image of human lungs off the internet. When I compared it to butterfly wings, I blinked hard, amazed at what I was seeing for the first time.

Human lungs are pretty much the same shape as a pair of butterfly wings.

Did observations of nature lead to ancient origami paper-folding? To an understanding of the folded structures of DNA proteins? To radiation shields designed for spaceships, which unfold when deployed?

Wings & Lungs

 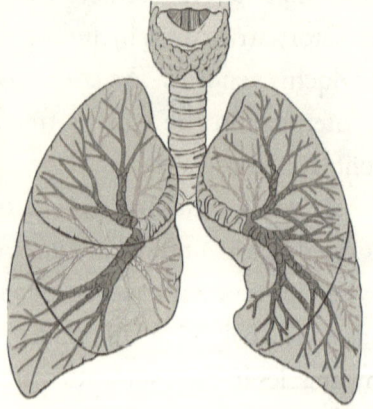

Afternoon

By midday, the sun was as high as early January allowed and had burned off the morning's low clouds. Taking advantage of a warm respite from winter, a neighbor met me to clear an area of invasive ivy and overgrown shrubs just outside the gates of our community's swimming pool.

"Are you going to release the monarch outdoors?" she asked.

"I haven't decided yet."

When my back, arms, and legs began to ache after several hours of weeding, I excused myself and walked the short distance home to check on my newly emerged butterfly.

I was happy to see it standing upright on the twig a few inches from the chrysalis shell from which it had emerged. Its wings were folded together over its back, showing their paler undersides. Long ago—way back in the 20th century—my earliest experiences as a student nurse caring for patients recovering from spinal cord injuries taught me to recognize, acknowledge, and celebrate such similar bits of progress.

Cartoon

Noticing my approach, the butterfly opened its wings wide, as if to show off their bright, beautiful colors and reveal their secret. An eighth-inch, round spot in a black vein on each of the hind wings confirmed my tiny companion's sex. The identification I'd made from his caterpillar markings held. *It* was now *he*.

As distinguished from sex, gender is probably not as complex among insects as humans. Although it is easier in spoken language than the passé "s/he" and much is gained by inclusivity, for me something personal is lacking in the newer, non-binary "they." Perhaps other solutions will emerge as language evolves.

It's a boy!

Evening

In English, we would never refer to a member of our family, or indeed to any person, as "it." That would be a profound act of disrespect. It robs a person of selfhood and kinship, reducing a person to a mere thing. So it is that in Potawatomi and most other indigenous languages, we use the same words to address the living world as we use for our family. Because they are our family.

—Robin Wall Kimmerer *Braiding Sweetgrass*[3]

In *Braiding Sweetgrass,* Kimmerer discussed calling non-human beings by their gender pronouns. Missing from English, this is an integral part of Native American languages. She called it the grammar of animacy.[4] Naming personalizes relationships between humans and others. "Dog" is a category. "Fido" is our best friend.

In that moment of recognizing his beautiful aliveness, confirming his sex, and noting with relief that he was healthy enough to move independently, I decided to name my housemate. I phoned my 3- and 6-year-old grandsons for assistance.

"Would you like to help me name him?" I asked.

When they stayed with me during the previous autumn months of COVID-19, both boys discovered monarchs outdoors, in various stages of their natural lifecycle—tiny eggs, caterpillars in their enlarging instars, full-grown caterpillars, and butterflies. They even witnessed predation by a sewing spider (*Argiope aurantia*) that had cleverly spun its web on my house a few feet above the milkweed plants. The boys were sad but curious when they saw that one of the monarchs had been caught and carefully wrapped, identifiable by the orange and black wings that showed through the Argiope's silk. We talked about it.

"The circle of life," we sang in unison.

But they hadn't seen any chrysalides until one of our recent, weekly talks. They were eager to see it as a brand-new butterfly.

"Let's FaceTime," the younger one exclaimed—the application's name had become a verb. We hung up our cell phones and switched to a visual call. Although he was not yet able to fly, the butterfly stood on a twig and fluttered his wings as if to show them off to the boys.

"What should we call him?" I asked.

"What about Fluttery?" the younger one suggested, noticing its wing movements.

"How about Spotty?" asked his older brother. "For the white spots on its head and wings."

"Great ideas. How about we call him Spot for short?"

We agreed.

"I name you Spot!" I said.

I felt like the Queen of England conferring knighthood.

My relationship with Spot instantly felt more personal. I worried about him, spent more time with him, looked at him fondly, and watched him, my eyes softened with feelings of kinship.

From the animacy distinction, my thoughts wandered to the 2018-2021 updated production of the futuristic "Lost in Space."[5] The child, Will Robinson, referred to the advanced artificial intelligence (AI) character as "he." In animals, gender differences are influenced by biological hormones, and are not yet built into robots. Different voices are available and draw us into closer relationships with our AIs. I like my GPS to speak to me with a typically male, British accent. Sometimes I talk back to him.

In the TV show, other characters saw only a machine they referred to as "it." While Will responded warmly to the characteristics we call human—compassion and thoughtful concern for Will's wellbeing— others assumed the AI's behaviors were merely advanced programming. They feared danger and kept their emotional distance. We have a choice to treat the unfamiliar with hostility or with compassion and science.

The advantage of AI is that much more data can be processed simultaneously using quantum mathematics. For example, computers can make complex, simultaneous connections among chemical processes that can lead to new medical therapies and block processes that run amok.[6]

After I'd been reading for several hours, I went to my closet for a sweater. The day's warmth yielded to a wintery drop in temperature. Cloud cover brought freezing rain. If I'd let Spot out earlier, he'd have died by this evening. Nature or nurture? Death or life?

Although I am decidedly pro-choice in most matters, my heritage suggests, "Choose life."[7] Adults should make decisions that support their health. At seventy, I choose to eat rather than fast on Yom Kippur. I feel better, can review the past year, and decide what I might improve.

Rationally, the success of my involvement in Spot's eclosing and my natural curiosity and hubris gave me confidence that I could figure out how to keep him alive. I enjoyed learning from his presence. I became emotionally attached as his willing servant and caretaker. I'd keep him indoors and do my best. As a result, I needed to feed him. But what?

Spot did not eclose under environmental conditions typical for his kin. Winter provided no natural nectar sources to bring indoors. I vaguely remembered having heard something about providing Gatorade. In 2012, I studied to become a Tidewater Certified Master Naturalist. It was mentioned in one of their classes, in the context of feeding butterflies in the Norfolk Botanical Garden's (NBG) butterfly house.

Postponing other plans, I drove to a nearby grocery store. So many enticing varieties were displayed. Which should I get? Red, yellow, or green? Fruit punch, lemon, or lime? Regular or extra strength? None were free of "added colorant," whatever that was. Since it was thought that red colors attracted hummingbirds to backyard feeders, maybe that would attract Spot. I decided to try it. Once home, I poured red, fruit punch-flavored Gatorade over all the tiny pockets of an organic sponge until it pooled on the saucer beneath.

How would I get Spot from his twig perch to that Gatorade-soaked sponge? Could I coax him to join me, rather than picking him up by grasping his folded wings and risking damaging them?

During that fall visit, my older grandson watched intently to learn how butterflies and dragonflies perched on twigs. Experimenting, he'd wait patiently, standing so quietly that he was barely breathing, with one arm extended with a pointed index finger. His efforts paid off when he graduated from enticing damselflies to monarch butterflies.

They rewarded his practice by alighting on his finger, one at a time. He'd watch, motionless, until they flew off.

Remembering his successes, I touched my extended index finger to the front of Spot's forelegs until he stepped up onto it—our first communication. Even human communication is mostly nonverbal. How did Spot know to do that?

Fifty years ago, when I was in nursing school, I learned that newborn human babies have a stepping reflex. To elicit it, textbooks and films demonstrate supporting a newborn upright, then touching the front of one foot to a firm surface, like I'd touched the front of Spot's leg. Newborns will lift that leg, then the other, while putting the first one down. It looks as if they are trying to walk. Another approach is to lay a newborn prone on the mother's belly, facing her head. An alert newborn will lift its head, look for, smell, or spot the mother's nipple, then flex and push off one leg, then the other, thereby inching toward its target, then suckle.

Get food. The survival strategies we call instincts are encoded in the DNA of all living things. Nature comes before nurture. Instinct is the spark that initiates a behavior. Rewards—in this case, satiety—spur organisms to repeat behaviors that support life. Life-threatening results lead to avoidance. Fatalities can result in extinction. Survivors evolve by adapting.

More complex animals, like human babies, need assistance to survive. It's a far stretch to relate this to viruses, which, without their own, use host DNA to reproduce, adapt, and evolve. David Quammen (1996)[8] described how some viruses become part of our cellular makeup. COVID-19 may cause unknown mutations. Through reading, observation, and experience, I learned how similar we are to all living things. All are subject to infestations.

Spot did not show the small or deformed wings typical of ophryocystis elektroscirrha (OE).[9] It is an unfortunately common, protozoan infection, which is highly contagious among monarchs. Infected monarchs and milkweed, if they host spores, should be destroyed if dilute bleaching is not possible. OE spores, if present on a monarch's abdomen, will stick to tape, and can be seen under an inexpensive, 100-power, handheld microscope or even a cell phone's magnification. If found, infected butterflies should be destroyed rather than released into the wild, which would further spread the disease. Spot remained indoors.

I carried Spot across the room, then turned my finger, destabilizing his perch, until he stepped off my finger and onto the Gatorade-soaked sponge. Although he should've been able to taste it with the sensory organs I knew were in his feet, Spot showed no response. Perhaps he didn't recognize Gatorade as food.

By ignoring it, Spot exhibited negative communication. To successfully nurture him, I'd have to observe, recognize, and respond to his positive and negative communication. Deep observation and response can benefit human relationships, too.

On Spot's second day of life, I was already beyond everything I'd learned in Master Naturalist training. The Gatorade didn't work. Searching further, I found many anecdotal references to feeding honey. Several suggested diluting it to one part of honey in six parts of water. Less water and it might be too viscous to suck up. More water and it might be too diluted to provide enough calories before his stomach was full.

Honey is a natural antiseptic sold today in purified form to aid wound healing. Norfolk's rescue team for stranded ocean animals used bee pollen to help sea turtles' wounds recover from boat strike injuries. Three-thousand-year-old honey was found perfectly preserved and covered with beeswax inside King Tutt's tomb in Egypt. Perhaps honey would help keep Spot healthy.

Nectar is mostly water but it contains other nutrients important to the survival of organisms that consume it (see Appendix B). When honey bees gather it, nectar mixes with enzymes in honey bees' saliva. Rather than being digested, it enters a specialized honey pouch. From there, it's passed to other bees, which pack it into their hives. More bees fan it to evaporate excess water before sealing it into the honeycomb with beeswax. The wax keeps it essentially sterile. In the absence of naturally occurring nectar, bees will eat some of their own honey.[10] Perhaps monarchs could use the concentrated nutrients in honey for energy.

To make it, I stirred one-half teaspoon of honey into one tablespoon of water, then put some on a clean sponge. Spot tapped his left foot into several of its full craters, seeming to recognize it. More reaction than to Gatorade. But nothing else happened.

Gatorade: 0. Honey water: 0.5—halfway there. Tasting was not the same as consuming. What would link the two?

I browsed the Internet using tactics I'd learned back in the twentieth century in universities' physical libraries. Using the logical operators defined by the nineteenth-century logician, George Boole (1815-1964),[11] I searched the internet for useful information by typing in "butterflies and food," then "honey or Gatorade and butterflies." When responses showed whole sentence options, I realized that more recent developments in artificial intelligence (AI) had updated the search process and made it easier to sort through the vast knowledge gained since Boole's time. Queries in focused, complete sentences lead to more specific information.

I changed my question to "How do butterflies feed?" Better answers came from more specific questioning. "What should I feed my monarch butterfly?"

Recent advances in computing power permit access to research and anecdotal information of varied quality. Studying requires information literacy and more focus in the absence of more time. The urgent matter was *how* Spot would drink since he seemed to recognize honey water as food. For their only nourishment, butterflies use their proboscis—a tubular, elongated mouth part like a straw that is uniquely adapted to take up nectar from tiny pools in flowers. It fascinated me to discover that the proboscis is anatomically analogous to an elephant's trunk.

I looked for—but couldn't find—Spot's proboscis.

> *He'll starve.*
> *Wait, don't panic. Look further. Think!*

When I grew fatigued from the day's thinking, I took a break. My thoughts wandered to other kinds of nourishment. Nature had given me a companion, a challenge, and another kind of nourishment—the intellectual kind.

I still followed the CDC's recommended restrictions to reduce the risk of catching and spreading COVID-19. To avoid crowds, I still ordered groceries online and had them delivered. I wore a mask over my nose and mouth during necessary outings. I followed the news intently, did research to get accurate background information, and wrote about it nearly every day. Learning about butterflies took my mind intermittently off COVID-19.

As an introvert, I didn't mind the isolation as much as the extroverts among my friends and family did. I met with them using visual platforms, including FaceTime or Zoom (kids, Mom, sis), WhatsApp (for friends in Argentina and Brazil), and WeChat (for friends in Asia). My geographically nearest, closest friends kept themselves as safe as I did. I met one or another occasionally for tea, a glass of wine, or rarely, a meal at one of our homes.

A few friends followed no guidelines. Fearing contagion or conflict, I didn't see them in person. I stopped trying even general conversations with COVID-19 risk deniers—all roads led to Rome. I judged them as selfish in their defiance and misled by the proliferation of unsubstantiated, inaccurate, and perhaps purposely provocative if not outright false news.

I also needed spiritual nourishment. I assuaged my grief for the hundreds of thousands of COVID-19-caused illnesses, deaths, and orphaned children by attending Shabbat services remotely most Friday evenings. Safely remote, I saw many people who'd been friends or acquaintances for over twenty years. Joining them, I lit a pair of traditional, small, white Shabbat candles, then blessed and drank some wine from the silver chalice I'd made in high school. We chanted traditional prayers of appreciation for the earth's grain and fruits that sustained us. Time slowed. Near the end of the hour, we said Kaddish, the life-affirming prayer contrastingly recited in memory of the departed. We focused on life even while grieving over death. Attending those Shabbat services was a gift. So was Spot. A tiny creature living in my home upon which I could focus my attention, interests, heart, and intellect.

Spiritual Nourishment

January 3
Morning

How will I get Spot to feed? What have I missed?

As usual, I awoke before six, warm and snug under a down comforter. Only my face was chilly, so it was good to hear the furnace fan's hum when it ticked on. Slowly, the indoor temperature rose above the 68-degree overnight setting. I remembered last night's concerns. Colder temperatures delayed eclosure in nature. Maybe Spot's maturation was similarly delayed. I felt driven to solve the feeding challenge and emerged from the blanket to take a hot shower and begin the day.

Sipping my morning cocoa, I resumed investigating how to care for Spot by searching online for articles related to butterfly nutrition. I discovered that a newly eclosed butterfly doesn't require food at first. The proboscis of a butterfly is among its last parts to mature. Sustenance comes from the DNA broth in the chrysalis. The pieces of the puzzle of how to nourish Spot were coming together.

It had not occurred to me previously that there were evolutionary relationships between humans and insects. Learning about a butterfly's delay in calorie intake reminded me of the feeding habits of human newborns I'd cared for during my nursing career.

Before birth, the human fetus' gastrointestinal tract is unnecessary, not functional, and plugged with meconium. Oxygen, carbon dioxide, and nutrients like glucose, protein, calcium, and metabolic wastes diffuse across the placenta between mother and fetus through the umbilical cord and placenta. The vascular pressure changes that accompany a newborn's initial hours of breathing adjust its entire circulation for independent life. The obsolete placenta and umbilicus separate from the uterus and are expelled, literally the afterbirth. This route of nutrition is eliminated.

Yet, breastfed newborns consume few calories during their first several days. Although parents' and grandparents' first questions include whether their little ones are getting enough to eat, most full-term newborns don't need forcing to suck. Early nursing provides antibody-rich colostrum and stimulates milk production.

Humans are born with special, calorie-dense, brown fat. It is located along the shoulders and between the shoulder blades. That brown fat is metabolized to provide calories until the mother's milk comes in. After several days, milk is sufficient in volume and calories to nourish the infant. What would a newly eclosed Monarch use for calories?

The question led me to more studying. I read about other structural analogs—in addition to lungs and wings—between humans and butterflies. Brown fat is also present in caterpillars. It persists in the soup of metamorphosis to provide fuel to newly-emerged butterflies while their wings dry and harden. It disappears once butterflies can fly to find nectaring flowers for food. This explains why the proboscis doesn't need to mature into a tube before eclose. Their brown fat is located on the thorax under their wings.[12] Shoulder blades…thorax… wings… brown fat. Maybe Spot wouldn't starve after all.

I considered another variable. Two weeks is the typical gestation for a monarch. Spot eclosed fourteen days after completing his caterpillar, pupa, and chrysalis stages. He had eclosed "full term," but what did that mean?

A mathematical formula to predict the expected due date (EDD) for the spontaneous birth of humans was discovered by Franz Karl Naegele (1778-1851)[13] who based his work on that of his contemporary and countryman, Carl Friedrich Gauss (1777-1855).[14] Gauss was a child prodigy and prolific mathematician who discovered a pattern to the distribution of naturally occurring phenomena.

Naegele was an obstetrician and professor of medicine and midwifery in Heidelberg, Germany. He used his records of hundreds of his patients' last normal menstrual period (LNMP), average cycle length, date of conception, and actual delivery dates to determine the EDD. To do this, identify the first day of the LNMP. Subtract three months. Add seven days. Advance the year if needed. Example: LNMP=June 21, 2023. EDD = March 28, 2024. This formula is still highly accurate and useful. Naegele showed that his patients' actual deliveries plotted out to a Gaussian or normal distribution, better known as a bell curve for its shape.[15]

Since there are increased risks of complications at the extremes, efforts are made to stop premature labor and deliver prolonged pregnancies. Babies born early have difficulties breathing, maintaining body heat, and digesting and often need supportive care provided by neonatal intensive care physicians and nurses. On the late extreme, the placenta—the fetal lifeline—degenerates and may lead to fetal distress.

This concept holds for the gestation of all species, including butterflies. Although the ambient temperature is an added variable for their incubation, Gauss and Naegele's work from the 19th century led to the discovery that the time from egg to instars to chrysalids to butterflies can be accurately predicted.

Evening

These first days of the new year are long. It is cold outdoors with few hours of daylight. COVID-19 continues to rage. I have lots of time for contemplation and a tiny creature that engages my attention. How can I apply what I'm learning to care for Spot?

I looked at Spot's head more closely than I had the day before, magnifying it with my cell phone's camera. With great relief, I found his proboscis tightly curled right under his face. Spot hadn't fed because it hadn't opened and fused yet. Perhaps he would when his digestive system matured a bit. There's so much to see when we take the time to look closely enough.

Tightly curled proboscis

Antonie Van Leuwenhoek (1632-1723)[16] came to mind. Looking through the magnifying lenses he was the first to create, he saw what no one could previously: microorganisms swimming in a tiny droplet of pond water. His discovery initiated a new era in healthcare.

I thought about the newborns I'd delivered. Sometimes they needed encouragement to start nursing. I'd learned to gently stroke a newborn's cheek to stimulate a rooting reflex. Stroking causes its face to turn toward the side stroked and open its mouth widely enough to grab onto a breast or bottle nipple, then suck and swallow. This survival instinct is deeply encoded in our DNA. Maybe Spot had a similar survival instinct I could stimulate so that he'd uncurl his proboscis.

I found an approach that seemed analogous to stimulating a baby's rooting reflex. Gently, carefully, I poked the tip of a toothpick into the open "O" in the middle of Spot's curled proboscis. He uncurled it to its full length and immediately prodded it into the sponge I'd soaked with the diluted honey. The tiny ripple I saw in that pocket of fluid told me he was sipping it. I'd accomplished something momentous and thrilling.

After drinking, Spot waved his proboscis all around for several moments like a cowboy practicing with a lariat. I learned another part of insect anatomy: he'd just discovered his basal galeal joint,[17] where the proboscis joins the head. Then he drank again all by himself without prompting. I was as mesmerized as if I were watching an infant discovering his toes.

With new energy from the nourishment, Spot practiced opening and closing his wings several times. Then, securely perched upright on his twig, Spot turned his head for the first time. He rotated and angled it as if trying out its range of motion.

I wondered how butterflies were able to move in the ways I'd just seen. It took me a typically roundabout way to learn that they have muscles substantially like ours. I learned that all animals have muscles that allow for movement between points of attachment. We can be agreeable by contracting and relaxing opposing muscles that pull the head up or down. We say "no" with muscles that pull side to side. Butterflies use their muscles to fly, suck nectar, and breathe through their abdominal spiracles.[18]

With two pairs of wings and all those legs, insects must have more muscles than the 600 in human beings. Even after hours of internet searching, I couldn't find out how many.

I read that insects use their eyes to identify predators, prey, and food. I looked at Spot's eyes and imagined all the muscles he used just to see. Human eyes each have only one lens inside. Insects have 12,000 to 17,000 tiny lenses called ommatidia in each of their compound eyes. Each of these ommatidia has a lens and system of nerves, making it possible for them to see multiple images—each from a slightly different perspective.[19] Their muscles and those lenses all work together to help them know where to land. They have nearly 360-degree vision even before moving their head to look elsewhere. I wondered if my small friend's initial head movements made him dizzy or fatigued.

Were there also parallels in the need for rest and sleep?

Yogananda (1893-1952)[20] was a yoga master who remembered being fully conscious at birth. He wrote about awakening at birth reincarnated into an unfamiliar body and described having to learn how to use it.

Human newborns are often fully alert right after birth. They recognize familiar voices and start to associate them with faces nearby as if asking, "OK, now I'm here, what's up?" After an hour or so—perhaps longer if the environment is subdued or shorter if there's overstimulation from bright light and noisy staff—they dive deeply into sleep as if to process all the new information. Their behavior reminds me of the fatigue I feel after some time at a loud party or after a couple of hours looking at paintings in an art museum.

How would a butterfly describe his first experiences moving all those muscles? Perhaps, while resting, Spot was thinking about his new form and marveling at his amazing capabilities. My attribution of anthropomorphic abilities showed how bonded I was becoming.

After a short time, Spot folded his wings up to conserve the warmth. The paler colors on the underneath side of his—and most butterflies'—wings would provide great camouflage. Spot entered his nighttime state of torpor. He rested—or slept—who knows? How would a scientist determine an insect's brain activity to differentiate between rest and sleep? Did they have alpha waves?

Scientists are still discovering what happens when humans sleep. Spent neurotransmitters and other biochemical wastes flush away. Maybe both newborn babies and butterflies reconnect with a spirit-life between physical manifestations beyond the veil between life and death. We'll probably never know.

January 4

Spot flies!

I happened to be watching. It was midday. The sun had warmed him and I had fed him. Unfortunately, right after he took off, gravity dropped him immediately to the floor. I couldn't tell if the fall injured him. Undaunted but exhausted, he climbed onto the forefinger I offered. I carried him back to his twig where he rested until evening. No prodding of his proboscis could entice him to unfurl it to eat.

Thinking about other cold-blooded animals like reptiles and amphibians, I was concerned that cooler nighttime temperatures might not be conducive to Spot's digestion. I remembered having short-lived pet turtles when I was a youngster. Cold-blooded like Spot, their metabolism slowed in the nightly cold temperature of my childhood home. They may have died from indigestion or pneumonia. In recent years, while volunteering with our local stranding team, we cared for Kemp's Ridley turtles received from Boston during a cold snap one Christmas. We had to warm them gradually. We weren't to feed them until antibiotics treated those with pneumonia.

I left Spot to rest, recuperate if he could, and integrate the day's experience into his learning. Anthropomorphic, yes, but why not? Isn't human behavior zoomorphic?

Spot became more active once warmed by the morning sunshine, which I supplemented with an incandescent lamp. He fluttered his wings tentatively then took off for a more successful, brief flight. After landing, he rested a bit before I led him to another sip of honey water. After a drink, he flew to perch in a nearby, sun-warmed window where he stayed completely still, wings up, for many hours.

When he woke up, Spot opened his proboscis without my prompting—a developmental milestone thrilling to witness. He waved it up and down, curled it tightly under his face, then unfurled and waved it around again. I felt lucky to be watching while it happened.

Was he exercising? Connecting neural pathways? Hopefully, he was preparing the appendage to take nourishment.

Then something unique happened. Spot moved it up over his head then *whack!* forward like a whip. After he repeated this maneuver a couple of times, a tiny droplet of fluid appeared at its tip. Then, *flick!* Spot whipped his proboscis a few more times. Another tiny droplet of fluid escaped from somewhere inside him. With repeated maneuvers, it slid a bit closer to the tip. After a few more flicks, the tiny drop reached the first one, combined with it, and doubled its size. He repeated this performance, whipping his proboscis up and down until a tiny bit more fluid came out. *Flick, flick, flick.* The droplet at the tip grew larger and larger until it could respectably be called a drop. One more flick and it flew off the tip only to land higher up his half-inch-long proboscis. Great shot!

Somehow, it stuck there. He flicked it repeatedly until the drop gradually moved by centrifugal force to the end. Then he shook it off. During the next fifteen minutes, he did it twice more. It was no accident.

Maybe Spot was clearing out the tube of his proboscis like a newborn will cough, splutter, and sneeze to clear out residual amniotic fluid from its airways. In both cases, a blockage is removed from similarly tubular structures to clear them to function properly.

I never saw Spot do that again. He had no trouble drinking afterward. Mission accomplished.

With basic functions assured, Spot moved on to other activities, motivated by internal commands. *Flap wings! Try a flight! Climb a twig! Wings up! Rest. Repeat. Wave proboscis! Look for food!* I was amazed, honored, gratified, and humbled to have such an intimate view of this tiny bit of wild nature.

The following morning, I was at Trader Joe's at seven when they opened. A brightly colored bunch of flowers looked like it would make a nice offering of gratitude. The daisies in the mix had a fresh, summery fragrance. I hoped he might respond to the scent and probe the bouquet for nectar.

I placed him on an orange daisy. He did nothing at all. He was indifferent, oblivious. He just stood there for the rest of the day. My gratitude went unnoticed. That was OK—it met my need to express gratitude for the opportunity to experience his life.

The next few times I checked on him, Spot had moved, first to the hibiscus and later to the window ledge. I put him back on the flowers I'd bought that morning, but he made no effort to probe them for a taste of any nectar they might contain. He displayed no curiosity that I could discern. His nonchalance showed me that he could tell they were out of nectar, out of season, out of place, probably hybrids, and not a recognizable form of food.

Spot on bouquet

Butterflies' antennae give them a significant sense of smell. Maybe a different fragrance would attract him. I sprayed on a bit of my favorite floral-scented cologne. Still no response. The scents combined to make cologne may have been from hybridized plants or whales and harvested for the perfume industry. Or perhaps attraction must be visual as it is for many human males. Another experiment ended. Spot would just have to drink diluted honey.

When I approached Spot to check on him before going to bed, he flew off. It seemed wonderfully normal and wild for him to shun human contact. He had passed his newborn stage. Spot was full of youthful vigor and in control of his body.

He fluttered up toward the ceiling but soon returned and approached my head. He endearingly landed on the back of my bright orange sweater, perhaps imprinting on me.

Flooded with the oxytocin of entrainment, I was in love.

When I tried to reach behind myself to have him perch on my finger, Spot flew again. This time, he managed a full flight around the living room, up and up toward the light fixture. Caught by surprise before I could turn off the slowly rotating fan above the lights, Spot got batted by its blades. Surviving this, he strategically found a safe perch between the blades and the lights and stayed put for the night. Lights out for both of us. I went to bed.

Nighttime perch on ceiling fixture

January 5

This would make a good children's story if it turns out well.

I looked for Spot first thing in the morning. He was not on the light fixture. He was not on the flowers. He was not by the window. I walked carefully to avoid the possibility of stepping on him. I found him under the buffet, lying on his side, very still.

Fearful of what I'd find, I kneeled for a closer look. An antenna waved slightly.

I practically ran to retrieve the fresh batch of honey water from the fridge, poured a bit into a small, water-bottle cap, and set it on the kitchen counter.

I went back to where he lay, crouched down, and put a finger by Spot's legs. He moved weakly, just enough to climb on. I carried him to the capful of honey water and set him close to it. He probed its surface with his front left tiny tarsus, the part of the leg farthest from the body— analogous to our foot. It must have been an acceptable taste because he stuck in his proboscis and drank for a long minute.

Tasting with his foot

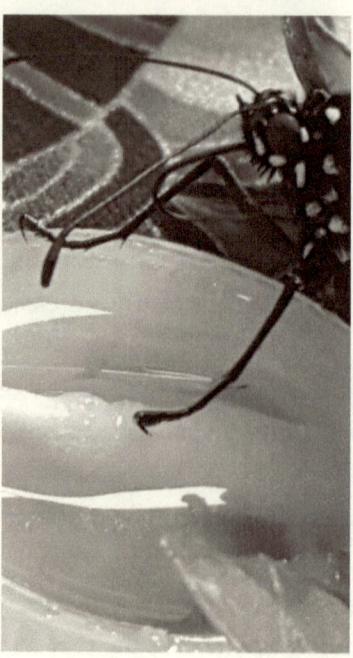

Humans can't breathe and swallow at the same time. Butterflies can. When we swallow, our epiglottis closes the pathway in the throat that leads to our lungs. If that failed, we would cough to expel any fluid that entered our lungs. Otherwise, we choke, gasp for air, and risk aspiration pneumonia and death. In contrast, butterflies respirate through spiracles, which are tiny openings along their abdomens.

Finally sated and revived, Spot opened and closed his wings. I swallowed hard with relief, then took a deep breath and slowly exhaled. Glad he'd survived another close call from death, I deduced that his collapse must have been caused by dehydration, hypoglycemia, or both.

I'd need to try to get him to feed more often, like a baby. That brown fat that sustained him earlier must've run out. I left him next to his bottle-capful of honey water and wondered if he might drink later, on his own. But as far as I could tell, he remained quiet with his wings up all morning.

When I checked back in the afternoon, he'd disappeared again. This time Spot was easier to find. He was on the narrow ledge of a west-facing window. But he was listing to one side.

Wondering what had happened this time, and trying to help him get upright, I saw that one of his feet was stuck or frozen between the caulking and the cold glass. He was listing because he'd been trying to free it, grown fatigued from the effort, and given up the struggle, wings up to conserve warmth. Worried I'd damage his foot, I grasped his whole leg with my index and forefinger covering above and below the joint to protect it and moved his leg very gently back and forth. Soon his foot came free. Another successful rescue.

I began to appreciate how vulnerable these tiny creatures were. Perhaps that's why they'd evolved to lay around 400 eggs—so enough would survive to save them from extinction.

It was cold next to the window, but that's where Spot liked to enjoy sunny days. To help him stay warm, I attached my task lamp to the table near the window. I placed his honey water on it and set him beside it. He siphoned up several droughts before curling up his proboscis.

Like a loving mother, I tried encouraging him to eat a little more, but he knew when he'd had enough and refused enticement. He climbed up the stick I'd left in the vase of flowers and rested close to the warm task light.

Curling up his proboscis indicated that he'd had all he wanted or could hold, which could be sufficient if he fed often enough. He had to learn how active he could be before requiring another meal, and how long the calories in the concentration of honey water I provided would suffice. Unlike a human baby, he couldn't cry to get my attention for more food, so we had to learn from each other. I had to make a consistent dilution, offer it frequently, and let him learn how much he needed. Like in humans, there must be properly functioning internal signals for being hungry, full, and sufficiently nourished for optimal health.

In nature, butterflies thrived when plenty of nectar was available for them to feed as often as needed, depending on the temperature and their level of activity. When it was over 70 degrees Fahrenheit, they were more active and needed more fluids, like us. Without the natural source of fluids, I had to learn Spot's needs and improvise.

A few hours later I checked again. What I saw made me sad. Spot had flown back to the sunny window, perched on its ledge, and looked toward the evergreen plants in my backyard. My first thought was that he could tell the difference between natural and artificial light. A lightbulb turned on in my head when I realized he had well-developed color and depth perception and perhaps a sense of where he preferred to settle. As a healthy male, he would have a strong sex drive, too. Was he pining for his lost freedom? Thinking to distract him, I led Spot to the honey water several times, but he wouldn't drink. I remembered again that if I'd let him out, he'd find no nectaring flowers or mates before he froze to death. Thus consoled, I let go of my non-productive rumination and left him by the window.

Pining for the outdoors

Could butterflies have emotions? In the human brain, the hippocampus is necessary for spatial memory. The amygdala forms new memories and links emotions to them. Will these be discovered in insects? If they remember where nectaring and host plants are located, might they not have emotions we haven't yet thought to discover? Or, rather than remembering, do they only sense location from chemical signals? These are questions for doctoral students, biologists, and entomologists. Possibly, they are just anthropomorphic fantasies.

January 6

It's the anniversary of my father's birth in 1917. I remember caring for him in 2013, from shortly before and until he died. During his final days, he was only drinking tiny sips of water. Disregarding the effort required, he used his walker, and with assistance took frequent trips to use the bathroom. It reminded me of the long walks that monarch caterpillars take before beginning their metamorphosis. The parallels inspired me to write "The Last Caterpillar."[21]

When I was a young child, my dad would solemnly help me bury pet turtles, fish, and birds after they died. We made a special place in the backyard of our house, which he'd built. After we wrapped the deceased tenderly in toilet tissue and placed it inside the cardboard roll, our funeral proceeded out the front door, down the stairs, and across the driveway. We'd turn left, climb three steps, and continue along the cement walkway toward the beach side of our home. The pet graveyard was near the entrance to the spacious yard. To the right of the walkway, beside the few steps down to the patio, and under shaggy juniper shrubs we'd made a small clearing. We dug there often enough that I feared we might disturb others' remains. I must have remembered the location of each tiny grave because we never encountered other skeletons. With little discussion and great tenderness, Dad showed me how to mark life's passing in ways befitting my youth. His attention helped me learn to take life's losses in stride.

My thoughts turned back to Spot. Why did he refuse to drink late yesterday? Had the honey water fermented to alcohol or become moldy? Was instinct telling him it had spoiled?

He was right where I'd left him. He'd survived another 24 hours. I turned on the incandescent light to help him warm up before checking the honey water. It smelled sour as I'd suspected. Observing Spot's behavior was helping me learn how to take care of him. I felt the weight of responsibility for his life. Captive, he had no way to take care of himself.

I made a fresh, two-ounce batch of honey water. That would make enough for several days. Keeping it in the fridge would retard its fermentation. I brought some to him in a little water bottle cap that I rinsed and refilled daily.

Perhaps thirsty from his overnight fast, Spot perked up quickly, recognizing the tiny container I'd placed right in front of him by smelling the fresh honey water with the sensors on his antennae. Placing one foot into the fluid gave enough of a taste for him to unfurl his proboscis, put it in the beverage like a straw, and drink. Suck, stop, wave, suck, stop, suck some more. After a few minutes, he fluttered off to land on the leaf stem of a nearby hibiscus houseplant. Once perched securely there, he shivered his wings for a few minutes but didn't fly. Perhaps he was doing this to warm himself after landing away from the light or to disperse the recently consumed sugars for energy throughout his body.

Drinking from a bottlecap

I yearned for a crystal ball and wondered how long he might live. I attempted to calculate his age following an analogy. We used to estimate one dog year equaled about seven human years.[22] What would be the ratio of butterfly days to human years?

In the wild, a butterfly might live two weeks before succumbing to illness, predation, adverse environmental conditions like the sleet I found him in, or just bad luck. In a protected enclosure like the NBG butterfly house, one might last a month. Under good conditions, an American man's life expectancy approaches eighty years. How many human years are equivalent to one butterfly day?

If eighty human years are equivalent to twenty-eight lucky butterfly days, one could calculate an equivalence or ratio. Dividing eighty by twenty-eight, one butterfly day is equivalent to 2.86 human years. Using this calculation, on Spot's fourth day, he was like an eleven to twelve-year-old human. About the time puberty starts. In nature, a butterfly would be looking to mate.

January 7

The ants go marching one by one, hurrah, hurrah!

Spot hadn't caught on to finding his food without my help, but tiny ants had discovered the capful of it that I'd left on the table.

I tracked a line of them back from their march toward the honey. Across the table, down its legs, along the floor, across it, and into the house from someplace behind the wall. I remembered a song from childhood.

They were coming indoors from under the baseboard. I wiped away the trail of ants like a deranged murderer. In the future, I'd keep the honey water in the fridge between feedings. A few more ants wandered in. They bunched up, confused at the sudden absence of their nest mates on the trail they were following. I killed them, too. Somehow, the message got back to its source quickly, either by the absence of scent from the food or the presence of a mysterious scent of death. I was only glad there were no more ants. I wondered if other insect predators might be hiding in my house. Outdoors, they were fine but indoors? Not so much.

Years before, I was one of the Master Naturalist volunteers inside the NBG butterfly house. Daily maintenance tasks included searching for and dispatching spiders that otherwise trapped butterflies in their webs for food. New butterfly house volunteers took spiders outdoors and relocated them far from the protected habitat. Once these volunteers saw previously beautiful specimens trapped and wrapped in webs, they killed offending spiders without mercy.

At home, I looked closely for spiders. I saw a few tiny ones hiding in the top corners of the living room windows and swiped them away with an ostrich feather duster while making a mental note to check for others every few days.

Sometimes, I wanted to keep Spot close by when I worked in my office at the other end of the living room. He always tested to make sure my finger was a suitable perch. He'd step aboard with one foot at a time as if doing a Tai Chi maneuver until he had all four in place. He stepped off the same careful way, wings halfway opened for balance, one foot at a time until he had a firm grasp on a milkweed stalk. His careful steps reminded me of a dear friend who had neuropathy from chemotherapy. It took them both a little longer to feel securely supported. With his hooked feet firmly anchored on my finger, I walked slowly to the office. Once Spot and I settled there, my dog came to nap close by. We had a companion.

Louie watching Spot

January 8

I must learn how often to encourage Spot to eat.

Morning light barely arrived. At seven it was gray and chilly. Spot wasn't under the ceiling fixture where he'd perched—dare I say slept—the night before. I stepped carefully until I found him on the living room carpet. He was upright but not moving. He'd probably run out of calories—again. I retrieved the small bottlecap of honey water from the fridge and while walking over to him, took the chill off it with the warmth of my palm. I wet a Q-tip in it and offered that to Spot by touching it to one of his feet. Spot immediately unfurled his proboscis and tapped it into the moist Q-tip. I led him by setting it in the honey water. Spot submerged his proboscis into the tiny cup and drank. In less than a minute, he began to open and close his wings, reviving quickly. With Spot on one finger and holding the bottlecap in my other hand, we moved slowly to the dining room table. I set him on a cloth napkin there, so his barbed feet would have something to grip.

I'd cut and sewn that napkin from a two-yard piece of fabric I purchased during a vacation to Japan in November 2019, only months before COVID-19 entered the USA. The fabric had symbolic images of clouds and dragons woven with golden threads on a deep blue background. His bright orange wings contrasted beautifully against the colors of the napkin. Spot rested there after his breakfast. In the absence of bright sunshine, Spot wasn't drawn to the windows.

By the time he would drink again, the gas fireplace warmed the gloomy evening. He extended his proboscis, which was nearly as thin as a human hair. When he finished drinking, Spot curled his proboscis tightly under his chin. He was finished for the day. I turned off the lights, left him on the cloth napkin, and went to bed.

Spot on cloth napkin with dragon

Before going to sleep, I thought about what I'd seen and looked for more information about how butterflies drink. I knew the proboscis functioned like a straw. Did butterflies drink by actively sucking or passively, by capillary action? I learned they did both.[23]

Capillary action happens in very thin, tubular structures. When the lower end of one is submerged in liquid, surface tension causes fluid to be drawn up and adhere to its interior surfaces. Capillary action is called adhesion. It's an attraction between the liquid and the tube. Adhesion is different from surface tension. We see surface tension, or cohesion when a container is filled slightly above its brim, and the molecules of water stick to each other.

A familiar example of adhesion occurs when a paper towel is touched to a spill. Some of it is drawn up into the towel. Water-based fluids like water and sap have this adhesive quality because of their low viscosity (thickness). Spring sap rises in trees by capillary action, permitting us to tap the thin fluid and boil it down to make higher-viscosity maple syrup. I figured that honey needed enough dilution for capillary action to occur, yet not so diluted as to provide insufficient calories to a full stomach. Animals must either consume enough calories at one time or frequently enough to prevent starvation. Lions and wolves eat large amounts at infrequent kills. Humans often eat three times a day, perhaps consuming more than needed to ingest enough appropriate nutrients among empty calories.

Back in the 20th century, before transdermal sensors were invented, nurses like me had to use three-inch long, very thin, glass capillary tubes to collect blood samples (to test for healthy glucose levels and state-required reporting) from newborns after making a tiny stab in their heels with a lance. Yes, they cried—it was important but awful. The parallel between those capillary tubes and the butterfly proboscis reminded me of how many inventions have come from observations of natural phenomena.

I found that butterflies use muscles inside their head to suck.[24] Capillary action brings nectar—or, in Spot's case, honey water—into the proboscis far enough to reach taste sensors there—in addition to those in their feet. Recognizing acceptable food, the sucking muscles take over. After meals, the proboscis is curled and tucked away for safekeeping.

Week 2

January 9

Spot is starting his second week of life. He's doing so well, and I've learned how to take care of him. I eagerly look for him in the morning, enjoy seeing what he'll do during the day, and wonder where he'll spend the night. Free time for this is one of the benefits of retirement.

It was a sunny morning. Spot breakfasted on freshly made honey water. Once sated, he flew wonderfully around the living room for a few minutes before grasping a window-shade cord with his four prominent legs. To Spot, that window-shade cord must have looked familiar, like a flower stem. He clung there the rest of the day.

Although I knew that most insects have six legs, I saw only four on Spot. To know why, I needed to understand how monarch butterflies fit into the taxonomy of arthropods. I read that in the subfamily, Danainae, which includes monarch butterflies, the two forelegs were often regressed and nearly hidden in the upper thorax. In some populations of male monarchs, they are absent (see Appendix C). I took, then enlarged, photos with my phone's camera to get a closer look. He had only four legs. I didn't know what that meant. Perhaps Spot was more evolved than others of his kind, or from a different sub-species.

The winter sun had set around five, but with artificial light and warmth provided by central heat supplemented by my fireplace, Spot was awake for enough hours to mimic summer's later daylight. He drank honey water around eight, then flew to the ceiling light fixture to perch for the night. He'd learned to avoid the slowly turning blades, which I'd once again forgotten to turn off.

How much time do butterflies spend flying in the wild? How much time do they spend at rest? Although I couldn't find anything definitive when I searched the internet, it seems that flight is dependent on ambient temperature, the age of the butterfly, the urge to mate, hunger, the availability of nectar, and the possibility of migration. The latter is only available to some of the last monarchs of the season. Triggers for the required longevity are under investigation. I didn't know whether Spot was predetermined for migration. If so, he'd missed the departure notice.

January 10

Can butterflies hear?

I was glad to find him hanging, wings down, under the ceiling fixture where he'd perched the night before. I used the rheostat to partially turn on the light, so he could gradually warm up. Once he flew to his customary, sunny window, it was time for our breakfasts. I fed my dog, then Spot, then myself.

"Just for fun, why don't you see if you can train Spot with the dinner bell?" my daughter suggested. She was referring to an antique, hand-cast bell I'd purchased in an outdoor bazaar in Tokyo.

I rang the bell and then offered honey water to Spot's foot, so he could taste it. He quickly unfurled his proboscis and drank. If I repeated this consistently enough, maybe he would become conditioned to anticipate his meals and do something to confirm it, like fly to me when I rang the bell.

Kathleen Lucas and her colleagues (2009)[25] identified the hearing apparatus in blue morpho butterflies. She determined they could distinguish between high- and low-pitched sounds and postulated this helped them detect approaching predators, such as birds.

Bell experiment

After breakfast, Spot briefly fluttered his wings, then took off. But he landed on the floor again. My early morning enthusiasm evaporated. I let him climb onto my finger. When he refused more drink, I took him to the window-shade pull cord he'd enjoyed the previous day.

Was the problem that he hadn't absorbed enough calories? Maybe the dilution must be more precise. Maybe he needs more honey in the mix. Or is it something else?

He was drinking as much as he wanted—or as much as his tiny body could hold. I offered it several times during the day; sometimes he wasn't interested at all. I became more careful in my measurements: exactly one part of honey to six parts of water.

Maybe he wasn't getting enough electrolytes.

I learned more about the history of Gatorade from my dear friend, Dr. Alan Bartel (z'l),[26] who, in 1965, was a medical resident and research participant in experiments conducted by Dr. Robert Cade at the

University of Florida. Dr. Cade wondered why their football team's initially strong performance declined markedly by halftime. His analysis of their pre- and post-exercise sweat showed that athletes lost a lot of salt, potassium, and glucose, which created their fatigue. Dr. Cade concocted a beverage to replace those electrolytes. He added lemon juice, at his wife's suggestion, which improved its taste. When the team drank it at halftime during its 1966 season, the Florida Gators greatly improved their second-half performance. They ended the series "with a combined total of 265-147."[27] They went on to defeat the Georgia Tech Yellow Jackets 27-12 in the 1967 Orange Bowl game with the help of what became known as Gator-aid. Gatorade.

I tried Gatorade again. I placed two clean, organic kitchen sponges side-by-side on a plate, and soaked one with Gatorade. The other got the 1:6 dilution of honey water. When I put Spot on the Gatorade-soaked sponge, he showed no interest in it—no foot-tapping to taste it, no proboscis unfurling to probe into it. I moved him to the sponge soaked with honey water.

As soon as his foot touched that sponge, it was clear he recognized something about it. He unfurled his proboscis and immediately tapped it into the honey water. He knew what was good for him or at least what was familiar. Wondering about the differences between the two fluids, I queried Google: "What minerals are in honey?"

Bingo. I found that honey has many more nutrients than Gatorade including calcium, copper, iron, magnesium, manganese, phosphorus, potassium, and zinc (see Appendix B). I wondered if Dr. Cade or any subsequent developers of so-called sports drinks thought of testing them against honey for energizing athletes.

Most evenings, I reviewed the day's events. My thoughts wandered through rational and emotional realms. I needed to escape from the emotional impact of all the unknowns around COVID-19. Spot's needs helped divert my thoughts during those dark winter days. His care and feeding needs led me to read more about butterflies.

How do butterflies feed in the wild? Just as honey is made from nectar collected from many flowers, when pollinators feed on diverse flowers, like when we eat a variety of plants, it's good for their health. Monarchs sip nectar from milkweed flowers and from others including penta, lantana, bee balm, phlox, aster, and myriad wildflowers when they are available. Although some of them look like they have one big flower, some consist of many clusters of little flowers, each of which contains nectar. Those are called composite flowers. Insects, including pollinators, favor them because they can just walk around and get lots of nectar and pollen. They don't need to burn the calories just consumed to fly to another flower. More bang for the buck.

Some composites, like penta and lantana, continue to bloom during the last weeks of summer and well into fall. They illustrate coevolution. The nectar from such flowers helps migratory monarchs store fat for their long journeys south to warmer climates.

Although I would have liked to give Spot more natural sustenance, it was too late to find any native flowers locally for him. Neither could I find any natives to purchase in bouquets. Spot would have to continue feeding on honey water. It made sense as a source of nutrition, and it was working—he drank it and revived. I just needed to make a fresh batch every few days, make sure its dilution was consistent, and offer it several times every day.

At times, I worried about how development, hybridizing, intensive agriculture, the raising of livestock, and the introduction of exotic species have disrupted our wildlife. Because of my little guest, I could redirect my musings from the enormous concerns and focus on this one creature and his family. I was often discouraged by thoughts that Spot might be among the last of his species. I struggled to keep focused but I got pulled back into the broader issues that confront our planetary survival. Although populations of monarchs appear to be stable despite environmental threats,[28] I can't help but wonder whether in my lifetime monarch butterflies could disappear—as have myriad species.[29] Butterflies might become another victim of the Anthropocene era of mass defaunation.[30] At a minimum, their iconic migration from northeastern North America to Mexico might end.

To distract myself from those depressing thoughts, I asked myself questions and looked for answers. Sometimes it was a deep dig. I liked connecting the dots between the discoveries of scientists through time to grasp my place in history. My life is but a tiny moment. There is context. Knowing that helps me discover the beauty in life. Understanding of the natural world grows over time as scientists use their minds to do research and mathematics beyond my understanding to learn about our world and share their findings with others. From that, perhaps we can do better to help ourselves and others, to live and let live. My musings would come to some conclusion and I'd come back to the matters in front of me.

How did the monarch get its proboscis? That could start a fable…

To get the whole picture, I read back in time. The emergence of flowering plants marked the beginning of the Cretaceous period about 130 million years ago (mya). Insects coevolved with flowering plants. As floral shapes diversified, so did the means to pollinate them. Insects evolved with structures that adapted to fit many floral shapes. The hair-thin proboscis of butterflies can probe even the smallest shapes and puddle into shallow groundwater for minerals. The result was that insects thrive, breed, and consume nourishment along with assuring reproduction of the plants with which they coevolved without pesticides.

Each species of butterfly has a specific host plant, which is the only species on which females lay their eggs. Spring arrivals from migration and overwintered chrysalids coincide with the proliferation of their host. Milkweed (Asclepius) is the host plant for Monarchs. Their caterpillar offspring only eat milkweed, which coevolved with them.

When I was a child, I didn't notice milkweed plants in the spring. But I enjoyed plucking fall's fat, bursting milkweed seed pods. I'd blow softly on the emerging seed and its attached, fluffy floss, then watch them drift away. It was autumn's equivalent to blowing on dandelion seed heads in spring. I learned more about milkweed sixty years after that childhood play.

Milkweed plants and the monarchs that eat it contain a bitter-tasting alkaloid chemical called carotenoid, which is toxic to many animals. Birds throw it up. Milkweed's adaptation protects both the milkweed plant and monarch caterpillars from being eaten by animals hapless enough to survive a first taste without learning to avoid it.

Milkweed has evolved to have gradually higher levels of toxins. Instead of being poisoned by the higher dose when they eat it, caterpillars and adult butterflies have continued to co-evolve to tolerate it.[31] With this adaptation, monarchs avoid many predators—except spiders, which have no adverse reaction to the butterflies and may be seen trapped in their webs. This is one of nature's ways of trying to preserve the monarch.

More problematic in this scenario than birds, which can fly to other places to avoid it, consuming even a small amount of milkweed can be fatal for cattle. It is toxic to them so ranchers, farmers, and dairies remove it to protect their herds. Round-up is sprayed to kill it. Seeds for food crops have been hybridized so they will not die from the pesticide. No matter how much milkweed we might plant in our gardens, it has been eradicated from vast grazing pastures, farmlands, feedlots, and fields where corn and hay are grown for livestock feed. The ongoing destruction of contiguous wild areas has increased the distance between wild food sources. Monarchs survive by moving to more favorable locations. Fortunately, milkweed's floss carries their seeds so they can grow elsewhere. When highways are widened, wildflower margins are at best mowed and, at worst eliminated and replaced with additional lanes of concrete. Traffic temporarily eases while contiguous habitats for non-human creatures are lost.

In "*The Song of the Dodo: Island Biogeography in an Age of Extinctions*" David Quammen (1996)[32] explained how significant gaps in contiguous habitats contribute to population extinction. His findings influenced groups like The Nature Conservancy to purchase properties adjacent to larger state and national parks. Expanding existing habitats is more effective at preserving species than adding small, disconnected pockets of nature, which help only the smallest animals including those that don't migrate. Insects generally require less territory than mammals, but migratory species of all kinds require significantly larger habitats than pastoral species. Although every little bit may help, our small gardens are merely islands that provide insufficient continuity between food sources and result in localized subspecies. Fortunately, some localities preserve wildlife corridors.

The propagation of non-native milkweed species, particularly tropical milkweed (Aesclepius curassavica), has an adverse effect on monarchs' migration.[33] When caterpillars eat it, the adults emerge with smaller, lighter wings.[34]

Species' needs are interconnected. Native plants and animals that co-evolve in time and space are mutually beneficial. Birds' eggs hatch in the season when caterpillars are prevalent. Baby birds need the protein that caterpillars provide to their diets to thrive, so parent birds collect them by the hundreds to feed their young. One could easily claim that the purpose of butterflies in the web of life is to provide caterpillars for baby birds. Prematurely warm temperatures can take these dependencies out of synchrony.

Hybridization and global travel long ago changed what we consider to be the natural world. Non-native flowering plants used for land-scaping include hybrids and exotics. People started hybridizing plants to bring out desired characteristics. There were, as often happens with human interventions, unintended consequences. For example, roses are bigger but few hybrids are fragrant.

The first known flowering plant hybrid was achieved in 1716-17 by a British gardener, Thomas Fairchild (c. 1667-1729).[35] He'd corresponded with Carl Linnaeus (1707-1778),[36] who was known for having developed a classification system for living plants and animals. Linnaeus taught Fairchild how to distinguish male from female plants so he could breed them by hand. Fairchild created many hybrids. This was highly controversial because Christendom believed all species were immutable and made by G-d at the time of Creation.

Gregor Mendel (1822-1884)[37] advanced foundational knowledge of inheritance when he intensively hybridized flowering peas. On the positive side, hybridizing led to our ability to grow more food and improve nutrition around the world than would have been thought possible during Mendel's life.

Imported plants, called exotics, evolved elsewhere. The soil in which they evolved may have contained different minerals. They may bloom earlier or later than native plants with which insects coevolved. Many hybrids don't attract pollinators at all. Non-native, invasive plants often take over native habitats and rob them of nutrients and water used by their predecessors. They diminish or replace plants bearing the fruits and seeds that evolved along with the creatures that need to eat them. The complexities of hybridizing and importing exotics challenge species that depend on native plants (see Appendix D).

I checked the milkweed seeds a friend sent me. They were over-wintering in my fridge and were safe and dry, but I had no idea what kind they were. I threw them out.

When I watched Spot, time slowed. The more I looked, the more I saw. Spot adjusted his four legs and stance a certain way before a flight. He readied his wings by shivering them for a second or two. Then, he adjusted their alignment, setting them just so, more upright than before. Then he took off. Up and up to the light fixture. Like how helicopters work. Perhaps Leonardo DaVinci watched butterflies before drawing his proto-helicopter long before real ones were manufactured. Spot fluttered around it for a few moments as if dancing with the fan blades, then settled into his favored upside-down perch for the night.

January 11

D's cat died. Send her a sympathy card.

Emailed greetings have mostly replaced personal notes. Anyone can find beautiful, animated e-cards, complete with appropriate music for any occasion. It's a great alternative and avoids paper waste and fifty-eight cents postage.

Even so, sending a handwritten note is an affordable luxury. Receiving one is something special. Birthdays, holidays, and significant events like the death of my friend's beloved cat suggested sending more than a quick email.

Butterflies are often used to symbolize the fleeting beauty of life. Notecards made with images of Spot could serve many purposes. A talented friend agreed to take digital photos. My sister volunteered to print my selections as notecards.

With the first several shutter clicks, Spot fluttered his wings, startled by the new sound. Soon he habituated and stopped reacting. I put him on one of the floral napkins I'd made from fabric purchased in Japan. He walked around, as close as he was going to get to real flowers. He tapped his foot on the fabric as if trying to taste the images printed there. He looked so beautiful, perhaps even content. B.'s photos turned out well. I gave him a set of the resulting note cards, and I send others on special occasions.

Notecard image Spot on crane and chrysanthemum cloth napkin

In mid-February, 2005, two New York artists known as Christo and Jeanne-Claude, husband and wife, opened an installation of 7500 monarch orange-bannered elevated gates they'd created on twenty-three miles of paths in Manhattan's Central Park.[38] My daughter and I went there to experience it as it was designed to be—walked through.

"There they are!" she exclaimed. The first flags were twenty yards ahead of us. As we got closer, we saw a few when we walked along one of the park's many paths. They were spaced to follow the curving asphalt up and over a bridge. We continued until they disappeared, stopped to look around, then walked some more.

"There—another one's over there!" she said. The more we proceeded through the park and looked, the more we saw. They appeared randomly placed. Watching Spot gave me another insight into that exhibit. Intercepting views of the orange gates in Central Park resembled following a monarch butterfly in nature on a summer day. A flickering spot of orange appears, disappears, and reappears farther on or perhaps off to the side. Maybe it's a different one. Then, there's another. The bright flashes of the orange flags offered a promise against the gray and brown, snow-brushed day. Summer, and its butterflies, would return.

"The Gates" installation in Central Park

January 12

Spot lives "free range."

When I set his breakfast on the floral napkin where he'd spent the night, Spot readily approached the capful of honey water and drank. He rested for a few minutes while his body converted it to energy. Then he maneuvered as before and helicoptered his takeoff.

Gracefully fluttering his tangerine wings, he flew back to the window shade cording. Although the fireplace warmed my living room, I switched on a spotlight I'd long had to enhance whites and shadows on still-life arrangements I'd paint. I angled it to shine an inch off the window to direct its heat, and he splayed his wings open to absorb it.

Spot was so beautiful and in his prime. Most likely his plump abdomen exuded pheromones and sperm. Had there been female monarchs outdoors, they certainly would have been attracted to him.

Two friends stopped by. Both wore face masks. They came to meet Spot.

"Oooh, he's beautiful. Does he fly all around the house?"

"Yes, he is 'free-range,'" I joked honestly. "But he usually stays in the living room." He'd defined a small territory. "Sometimes, I carry him to my front-room office. He's perched on my sweater a few times, and stays put when I walk around the house."

Do all animals have territories? Near our Florida condo, I'd often seen pairs of Zebra butterflies in one area and other pairs elsewhere. Snorkeling, I'd seen fish in their territories around specific coral heads. Food availability, energy expenditure, and safety from predators had to balance favorably against any nutritional benefits of travel. Although people had their homes and shopping routes, those privileged with more mobility may get food from practically anywhere with little effort. Even less since online shopping and grocery delivery have flourished, an entrepreneurial benefit of COVID-19. In wild nature, our planetary cohabitants must hunt for food. Carnivores hunt for animals. Herbivores hunt for plant materials, including nectar. Omnivores, like humans, can live with a variety of available food options.

How big is a wild monarch's natural territory? I'd only seen them locally. I didn't think about why I saw them in the few places I lived in the Midwest and along the East Coast. Most of what I'd heard about them had to do with their iconic migration. I learned more during my studies to become a Master Naturalist.

Groups like Journey North (journeynorth.org) and Monarch Watch (www.monarchwatch.org) have sponsored the tagging of many thousands of monarchs. Dr. Fred Urquhart (2011-2002) was a Canadian zoologist who studied monarch migration for over forty years. After thousands were tagged by professionals and citizen scientists, he and his wife, Norah Roden Urquhart (1918-2009) found the first tagged monarch ever to be noticed in Mexico by anyone who would know what the tags meant. It was from Minnesota. Journey North has continued the Urquharts' efforts. Orley R. "Chip" Taylor (1940-) founded Monarch Watch (monarchwatch.org) to further educate the public about the migratory habits of this species.

It's rare to find a previously tagged monarch to check on its origins. But often enough, tagged specimens are found dozens, hundreds, or almost two thousand miles away from where they're tagged. Enough of these have been found to identify territorial patterns and migratory pathways. Can they adapt to a reduced or varied nutritional habitat? They can—population studies show stable populations overall but with variations in their locations. Where there is tropical milkweed and temperatures remain above freezing, monarchs tend to stay in place. Otherwise, they travel and proliferate where there is fresh milkweed and more nectar. With hundreds of eggs laid by each female, only a few need to survive somewhere to repopulate.

January 13

Spot feeds several times a day but not in response to the bell.

For the past several days, I'd continued to ring the bell before offering honey water. Spot never responded to its ring with any discernable intention to feed. I was asking too much. Thomas Edison said that from failed experiments, he knew many things that didn't work. I ended the experiment.

A soft quiet filled my home during those wintry days. Had others, also isolated during COVID-19's virulence, experienced such silence? The hum of the heater's fan, the fountain atop the fish tank, an occasional crackle from the fireplace—it was so quiet my high-pitched tinnitus often joined the chorus.

During last winter's first COVID-19 lockdown, I visited my mother in Florida. It was so quiet that when I walked my dog in the evenings, I heard a new sound there—the chirping of thousands of small lizards living under the foliage widely bordering her condo buildings' broad foundation. Previously, their chatter had been drowned out by noise from street traffic. Their songs made the same magical chorus I'd heard years before when our small group of tourists floated silently at night on a small boat in the Brazilian Amazon. I'll never forget either experience.

January 14

Did all aerobic life evolve to love sunbathing?

Where Spot perched on the window ledge, bright sunlight shone through his wide-opened wings, making him look like a stained-glass ornament. B. had come back to see if he could get better photos than on the grey day of his previous visit. For an interesting contrast in shapes and colors, we tried to have Spot stand on the keys of the piano I'd inherited from my grandmother. Spot clung to the Q-tip I used to carry him to it. With each try, Spot flew right back to the window, seeming to prefer the sunshine. He wouldn't step onto the piano. The one image we got was undignified and kitschy. I never had it printed. After several of his escape flights, we gave up. He wasn't having any of it, which reminded me of my grandkids, who often refused to pose sweetly for the photos I wished to take.

Spot readily drank an extra, midday meal, then rested in the sunshine. After dark, I fed him on the floral napkin at my dining room table, where he spent the night. It took me much longer than it took

Spot to realize that the piano keys were too slippery for him to gain purchase.

Soaking up sunshine

January 15

Spot might like to perch on a small, rough tree branch I could place right next to the window. He could get the most sunlight there.

I made a new batch of honey water and dripped some on a branch to imitate sap, which is also high in sugar, salt, minerals, and amino acids (see Appendix B). Spot perched there and drank it, showing that he didn't require the tubular nectar receptacles his proboscis had evolved to fit. I didn't try actual sap or maple syrup.

On twig with "sap"

Week 3

January 16

There's another ant invasion!

A new infestation of ants arrived using different tactics from their predecessors. These marched single file from the top corner of Spot's window, down along its edge, and onto the branch I'd propped there. They found the dried remains of yesterday's sap-like honey water. Perhaps they'd find Spot and carry him off for lunch. Horrified, I quickly sponged the ants away, removed the branch, and washed off its sweet lure. I found a nearly invisible web in the window's upper left corner. A tiny, opportunistic spider was waiting there for the ants—or for Spot. What chemical receptors told carnivores where to forage? A mystery for another time.

I grabbed a long-handled ostrich-plume duster—a souvenir from South Africa—and wiped away all traces of spider web from the window. I washed the floor and dusted the blades and mount of Spot's favored ceiling fan—his nighttime perch. I checked every hour the rest of the day to kill straggler ants until some sort of chemical death memo got back to their nest outdoors.

Spot had learned his source of nourishment—bell or no bell. When I put the honey water-filled bottle cap in front of him and pushed it gently against his front legs, he immediately put his foot in for a taste. His right foot. It was always the same—I checked many photos. Who would have thought a butterfly would favor one side over the other like humans? That could make an interesting topic for a graduate student.

Having made sure it was fresh, he unfurled his proboscis. He tapped it quickly several times on the edge of the cap and got closer to its contents with each tap like a visually impaired person using a cane or reaching for the handle of a door. As soon as Spot found the fluid inside, he drank. I wondered if he was far-sighted like my dog.

Far-sightedness would be a practical, evolutionary adaptation. Survival and procreation would fail if flying butterflies relied on near-sightedness to find host plants, food, and mates. They are thought to be near-sighted, but have broader visual skills than humans, with nearly 360 degrees of vision. They can differentiate among a broader range of colors—including ultraviolet.[39] They respond to pheromones to find potential mates.

Ancient humans were hunters and gatherers, tasks which required different far- and near-sighted capabilities. Did hunters have more acute far-sightedness and gatherers more near-sightedness? Differences in vision have not been definitively related to societal roles.

Spot drank for about twenty seconds, rested, and then took a few more sips before curling his proboscis and tucking it away. His facial parts were so dark it was impossible to see what happened there with the naked eye.

After each meal, Spot would fly back to one of three window perches. He generally stayed in the brightest one until dark. Although the windows were old and chilly to the touch, Spot may have absorbed radiant heat. Would frosty nights stress Spot if he slept by the window?

I'd learned, from their untimely deaths to keep childhood pet turtles warm enough to digest. Like those small turtles, butterflies are ectothermic—cold-blooded—and depend on external sources of heat to survive. Between 270 and 300 million years ago, warm-blooded animals diverged from cold-blooded ones. I enjoyed looking things up in our family's Encyclopedia Britannica as a child. Today, we search for information on the internet. I taught information literacy to many students and required them to look up the original research behind health-related news reports. Most often, I use Google's search engine. For an unusual view of evolutionary relationships, I found OneZoom.[40] It has an informative, interactive, fiddle-head-like tree of life.

I provided Spot's evening meal on my dining room table away from the window, so he could digest before the thermostat lowered the indoor temperature. Otherwise, undigested food might have fermented in his gut.

January 17

Spot is eating less often.

Not thirsty? Not warm enough? Too sleepy? Was he missing some minerals or micronutrients that he would otherwise find in nature? Were these signs of aging? I looked for a fountain of youth for him and perhaps for myself.

I tried moistening some of the soil from my worm bin to see if Spot might be interested in sucking up its minerals. No response except, I projected, "Yuck!" Maybe something else was wrong. Maybe nothing was wrong.

January 18

Spot takes longer rest periods.

I recalculated. 1 butterfly day = 2.9 human years. 16 butterfly days = 46.4 human years. Was Spot in his prime of life or middle age? When I was 46, I realized my intense work schedule took more out of me than it had when I was younger. I had to rest more on weekends than in previous years to prepare to teach the next week's courses.

Sunlight that radiated through the window where he perched still made Spot's wings glow as if made from stained glass. But the formerly velvety blackness of his wing margins was fading, and small cracks were forming. His time was passing.

Spot glows like stained glass

January 19

Sometimes, Spot flutters wildly against the sunny window.

A desperate attempt to escape? Depressed about his captivity? I saw his reflection there, fluttering back at him. Perhaps he thought it was another monarch and was optimistically showing off his best mating dance. Although most female animals don't live beyond their reproductive capability, viable sperm production continues throughout the lives of males. Perhaps he was entering middle age. Some efforts that used to be easy require more recovery time.

Fatigued from the extreme fluttering, Spot was lying on his side in the window when I found him later in the day. He revived briefly after he drank, then didn't move again for several hours.

Spot fed only twice most days now, which I accepted as a result of his aging and decreased activity. I disturbed him with offers less often. His feet didn't seem able to support him as they had. I placed him between the folds of a soft, paper napkin for support.

Wild Fluttering

Since it was a chilly day, I turned on the fireplace for more heat for Spot. I remembered that when my parents aged, they were usually cold—unless they were in Florida. Why migrate further? Spot was still as beautiful as ever but didn't seem to take as much of an interest in activities he'd enjoyed such a short time ago. No flying off, no night-time spent under the ceiling fan. I saw no new signs of deterioration, no more ants or spiders, and no signs of parasites. I left him for the night on one of the Japanese-printed floral napkins he seemed to like. I started to prepare myself emotionally for his inevitable demise.

January 20

I am relieved that Biden will be sworn in as president today.

I found Spot where I'd left him on the dining room table. His wings were folded tightly to conserve heat. One foot was on the honey water-soaked Q-tip I'd left there. Perhaps he'd taken a drink on his own. I lit the fireplace and put on the overhead light to help him warm up. He took a long drink from a fresh batch of honey water.

With full attention turned to the television, I watched the new president being sworn into office. The former president left the White House. I watched to see the transition of leadership with my own eyes after the previous month's election-related drama. I was glad to see the fundamental element of democracy—voting—ruled the day. We also gained our first female vice president.

Kamala Harris is VP

Joe Biden's swearing in

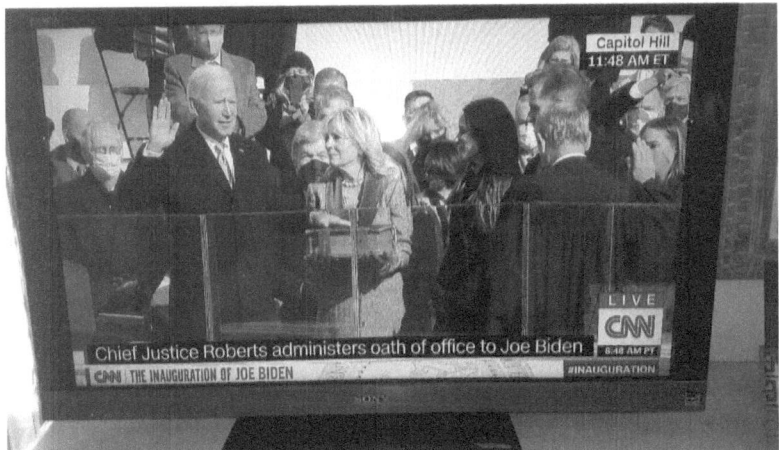

Week 4

January 23

I have competing obligations. How will Spot adjust to missing his morning feeding?

A few weeks ago, I joined the Virginia Beach Medical Reserves Corps (MRC) and soon completed the prerequisite online courses required to give COVID-19 vaccinations. The Board of Nursing approved recently retired nurses returning to service for this purpose. I scheduled myself for clinics on Mondays, Wednesdays, and Fridays, from six a.m. to one p.m. I planned to take food and water and wear warm layers of clothing under my MRC volunteer shirt. Vaccine clinics would be held in the convention center, a fifteen-minute drive. Spot would have to wait to be fed until I got home.

January 25

This is my first day giving COVID-19 vaccines.

Spot hadn't yet moved in the cold morning when I left before dawn for my first shift. A diverse crowd lined up outside the convention center. It was dark. It would be another hour before the doors would open. Still, at least fifty people stood alone, spaced themselves for safety, or in couples, their breath clouding the frosty air. A few held a small child in their arms. Volunteers were allowed into the lobby first.

When about forty of us had signed in, we were directed to pass through a single doorway to meet for instructions. Being of short stature, my vision was limited by the squeeze of the crowd. My heart raced. Passing through it felt like childbirth. Once freed from the narrow passage, my breath caught in my chest at being in an even larger crowd with all the organizers, assistants, and city employees assigned to help—their workplaces had been closed for months as a precaution. It was shocking to see more people in one place than I'd seen in over a year. I imagined everyone felt similarly.

After the team meeting, each of us went to our assigned table and got busy. I had no more time to dwell on the crowd. A variety of supplies had already been placed in a plastic box on each of the fifty cafeteria-style tables. At each one, the assigned vaccinator instinctively did the same first thing—grabbed a wet wipe to sanitize the tabletop even though they looked clean. A force born of habit.

After scrubbing mine, I opened the plastic supplies box and pulled out instructions, alcohol prep pads, Band-Aids, pens, cards to write the serial and lot numbers, and the date and location to give out with pre-typed instructions, and a clipboard. There were also forms for manually entering data, a useful redundancy in case the software-loaded laptops—provided to us to log vaccines—crashed. Each vaccinator worked with an assistant. We gathered again to receive verbal instructions on how to use the health department's software, recent updates, which type of vaccines would be given, and what information to provide for getting boosters.

The doors opened. People were directed to come in, single file. When they entered the large lobby and saw the crowds, most stopped like deer in headlights, shocked and scared to see hundreds of others. Like us, they had been quarantined for most of a year. Everyone wore masks. Providers like me also wore identification shirts and clear plastic face shields, gloves, and booties as carefully as if prepared to assist surgeons. We engaged one person at a time, showing the crinkles beside our eyes that indicated smiles beneath our masks. We did our best to set attendees at ease. It was personally rewarding to join the efforts of dozens of volunteers, first responders, city workers, medics, pharmacists, computer pros, and runners to quell the pandemic. As the morning progressed, everyone did their job. Every task was important. I was impressed that the firefighters, assigned to help, cheerfully emptied the trash when needed, which was often. I probably gave thirty vaccines that first morning.

When I got home, I tossed my coat aside and looked for Spot. He had found his way to a sunny window and was fine. As soon as I offered my finger, he climbed up on it, unfurled his proboscis, and began probing for food. No need for a bell—his response was conditioned by my feeding him. Feeling guilty about my tardiness for his morning meal, I quickly took him to the dining room table for a sip of yesterday's freshly made honey water. He sipped for nearly three minutes, the longest time yet. He was still strong: he'd been hungry but hadn't passed out.

Vaccination clinic at the Virginia Beach Convention Center

January 27

Taking care of Spot anchors me. I must be present, mindful, and focused to attend to him. I enjoy the calm, quiet rhythms of my winged guest's daily life, especially during the pandemic when so much is uncertain, unknown, and anxiety-producing.

During my second six-hour shift helping with vaccinations today, I spoke with hundreds of people as did each vaccinator. We spoke to those getting vaccines, their accompanists, friends, parents, other staff, volunteers, pharmacists, runners, firefighters, computer specialists, and managers. I gave about forty vaccines and got home at about two p.m. Exhausted, I showered and napped for several hours. But not until after I took care of Spot.

After feeding, Spot flew to a different window than usual—one that faced north. Previously unnoticed, its brightness indicated the darkest winter days had passed. The sun was moving north. Noticeably longer days were returning. From there, Spot fluttered to the top of the living room skylight for the first time. I stood there, stunned by watching him—those were magical moments. When he was tired, unable to gain a footing on its smooth surface, he returned to perch on my Ficus tree next to that north-facing, sunny window.

Spot on Ficus tree with Star

January 28

Spot is sensitive to temperature.

Falling from steely skies, fat flakes accumulated atop the morning's inch of fresh snow. Lacey ice crystals decorated the window next to Spot's nighttime perch. He stepped stiffly onto my finger, and I carried him toward the dining room table and its white cloth covering. Warmed after a few minutes away from the window, he turned his head this way and that. He was stretching, stiff from the overnight chill. Once warmed, he drank his breakfast, then curled and tucked away his proboscis.

Lacey tablecloth

With honey water dried there, why didn't it get stuck in that curled position? I couldn't distinguish among his face's black, tiny parts. To solve this, I used the zoom feature on my phone's camera to see better. For the first time, I watched Spot clean his proboscis after eating. He wiped it on his foreleg like an untutored child, then coiled and nearly hid it below his face. He must've found it still sticky. He uncurled it, wiped it again, and curled it up. Then again: uncurl, wipe, curl, repeat until, after several such sequences, he was satisfied it wouldn't stick to itself, he curled and tucked it away. No napkin needed, Mom.

January 29

I often talk to Spot.

As the month passed, I grew more attached. The longer I cared for him, the more I wanted to learn about him, like a lover who desires to grow familiar with every inch of his beloved. By enlarging the photos I'd taken, I noticed more details of Spot's head and thorax than I could see unaided. I became curious about everything I saw—and about what I couldn't. I found answers to most of my questions in scientific papers posted on the research-focused website, The Thoughtful Monarch [Facebook group],[41] and in Chapman (op. cit. 2013). I read, looked at Spot, read more, and repeated this—often.

I was reminded that the major functions of all insects are bundled into three tiny, oval packages stuck together invisibly, end-to-end. Head for instincts and feeding, thorax for breathing and mobility, and abdomen for digestion, reproduction, and elimination. From drawings of their insides, I imagined the invisible: minute nerves and circulatory vessels running inside and between the joined segments. Antennae were the easiest to examine in life.

Like a person who might raise one eyebrow at a time, a talent that I lacked, Spot often moved his antennae independently of each other, one this way, one that way, or only one at a time. Their far ends were enlarged and set at an angle, more like a hockey stick than a golf club. This is a characteristic of all monarchs. Their antennae detect specks of airborne chemicals from a wide area while they fly or that waft by when they are perched. They detect the fragrances of flowers and the species-matching pheromones of prospective mates. A butterfly won't try to mate with a bee.

Fascinated, I dug deeper. Antennae—informally, antennas—the name was first used in English in the 1640s to describe something the appendage unique to insects. Specialized chemical receptor cells in monarch antennae identify milkweed plants they require for egg-laying. It is the only plant their caterpillars can eat. Different butterflies identify their species-specific host plants and preferred nectar sources. The nob end is the primary functional part with most of the rest of the antennae carrying received messages to the brain.

Adopted from its Latin and earlier Greek origins, the name was used as early as 1902 to describe aerial wires used to send or receive manufactured, electronic signals. In my childhood in the 1950s, people placed a V-shaped pair of antennae, attached to a saucer-sized round base, atop early television sets to improve the black-and-white image clarity. With today's wireless technology, such antennae are relics for theatrical sets.

Below his antennae, Spot's oval, compound eyes—larger than ours in proportion to the rest of the face seen under magnification—looked like whorls of sunflower seeds on ripe heads. They occupied most of his head segment. Similar in overall size to the eyes, the curled proboscis underneath them functioned as his mouth.

Behind the head, the front of his thorax was hidden behind something that looked like tiny feathers. They were black with white spots. Along with the spots on his wings, they gave my grandkids and me the idea for his name. Spot's legs originated from the underside of his middle segment or thorax. I saw his upper and lower legs and feet with joints between each part something like ours.

I looked at photos and diagrams to better see what I tried to find on Spot. Tiny muscles attach to insects' exoskeletons since they don't have bones. The upper leg segment joins the body with a non-mobile joint. Even magnified, the point of attachment was hidden by the black blur of Spot's body. Following their length, I saw graceful bulges of muscles on his upper legs.

Muscles control the upper leg movements and are analogous to our quadriceps and hamstrings. The upper leg joins the lower leg by a hinge joint and muscle, almost like a human knee. The lower leg is of about equal length but thinner than the upper. A few small, hook-like sensory 'feathers' are located off the back of the rod-straight, lower leg.

Monarch butterflies have a pronounced barb underneath the lower end of the leg where it meets the ankle and foot sections. Joints were invisible but evident by the movement they enabled. Several feathery barbs are located under their ankles and feet, are sensitive to pressure, help the butterfly locate perches, and make such beautiful landings. A thicker muscle is attached to a pronounced, backward-facing barb at the very bottom of each foot. This barb makes a hook to tightly grasp whatever they've landed on, so butterflies can effortlessly rest at any angle without risk of falling off.

Spot's grasp was so strong that it would take some time and conscious effort for him to dislodge himself from one location before moving to another. When the window's icy glass numbed his feet, it took more neurological time for the idea and impulse to translate into action.

I'd learned a lot about insects in general and monarch butterflies specifically. More pressing issues demanded attention. The time I needed to get trained, staff the vaccine clinics, and recover from the intense work there drew me away from Spot in a good way. To set a patient at ease enough to receive a new vaccine from an unknown stranger took a nurse's ability to connect with honest caring in an instant. Working with others broke up the isolated days. There was so much to learn about this virus. I was grateful for all the virologists, epidemiologists, physicians, and nurses who struggled to understand more about it every day. In the broadest sense, for every subject, there is more to know than any one person might learn in a lifetime of study. Even narrowly focused scientists appreciate that knowledge evolves. Those thoughts comforted me and enabled me to focus on addressing the requirements of each present moment.

Week 5

January 30

It's been four weeks now, 28 days. I think Spot has reached the outer limit of a captive butterfly's lifespan. His wings are brittle. A few tiny scales from the tips of his forewings have broken off.

I reassured myself that breakage happened in the wild due to normal aging, drying, and crashing into branches. Spot's aging was not solely a product of captivity. I recalculated.

28 days x 2.86 human years/butterfly day = 80 human years.

Spot was beginning his old age. His life was passing quickly. If the weather hadn't been so cold and rainy, I might have considered letting him fly free one last time, thinking he might not be able to fly much longer.

When he was 83, my dad suffered a severe bruise from water skiing. Thinking how I'd never been able to do it, I was ungraciously amused when he commented with a sad tone in his voice I'd rarely heard.

"I guess I won't be water skiing anymore."

January 31

Spot spent the night on my dining room table. His wings are in tatters. He can only fly short distances now. Sometimes he won't take off at all. He's stopped flying to the ceiling light fixture at night.

More scales from Spot's wings broke off every day. His wings were splitting along their length like damaged fingernails. Was it possible to splint butterfly wings? An internet search offered conflicting advice with pros and cons. Nurture? Use very lightweight, small splints. Nature? Let nature take its course. Would it be more humane to put him in the freezer so he could die as he would in winter's cold? Wasn't "humane" anthropomorphic?

My emotions were involved, and I needed consolation. His days of exuberant living were dwindling due to the deterioration of his powers of flight. Yet he was still able to get nourishment and seemed to enjoy, if butterflies felt joy, spending his days warmed by the fireplace or perched in the natural light of bright, if gray, winter days. I decided to wait. Deciding not to do something was still a decision.

Warming by the fireplace. Tattered wings

Part 2

February

The butterfly effect:
A small change, such as the wind
created from the movement of a butterfly's wings
may initiate a larger effect later on
and elsewhere, even a tornado.

—Edward Norton Lorenz (1972)[42]

February 2

Spot can't fly anymore because of the tears in his wings. How will he adjust to this altered state? He's already adapted to limited mobility. Is "adapting" anthropomorphic? He is still alive and eating. Is that enough to let him carry on? Should I try to fix his wings? Or should I put him in the freezer to gradually go to sleep and die? Is it time?

Succumbing again to nurture, I read instructions posted online about repairing wing tears. I had to try. Left-brain intellectual curiosity won out over a right-brain emotional reverie. In the way of all scientists, including dilettantes like myself, I was curious. Perhaps it would be at his expense. I tried to avoid ascribing human characteristics.

His wings had so many tears that I'd have to make large splints that might make his wings too heavy for him to fly. Neurologically speaking, butterfly wings were like fingernails, without nerves, so the process wouldn't hurt.

I assembled supplies. Local hardware stores no longer carried the contact cement repairs suggested online. Rummaging through my toolbox, I found multi-purpose white glue, which would have to suffice. I assembled a hanger to place over Spot's wings near his body to immobilize him for the operation, glue, toothpicks, an index card to trim for the splints, and tweezers to help apply them.

Following suggestions I encountered online, I traced tiny replacement pieces to overlap and reinforce the broken areas of his wings. I trimmed them, so they'd be as small as possible. But to reinforce all the breaks, the splints would cover more than a third of his forewings.

It was time. I took a deep breath and immobilized Spot on the dining room table. I placed the right-wing splint first. Immediately, Spot tried it out. Fortunately, he could lift it. The effort tired him out and he stopped. I applied the left-wing splint and then released him. He flapped his new wings. He'd have to learn to coordinate them enough to fly. Placed near the window, he opened and closed his reinforced wings a few times before he stopped for the day. I fed him in the evening, and he spent the night perched where I'd left him.

Wing Splints

February 3

The splints are a disaster.

By midday, it was clear the splints had failed. Spot did well in the morning and appeared to gain the strength required to move his new wings. But when I checked on him after lunch, the wings had completely broken at the splint margins. The card stock was too heavy for his tiny muscles to support. I should have tried small strips of writing paper. I felt horrible about it.

With sad mercy, I cut off the splints where they had broken his wings. Although his hind wings were intact, only about a third of his natural forewings remained. It was not enough for him to fly.

Free of the weight that modulated their normal frequency, the remains of his wings fluttered at three times their usual speed. His forewing stumps and hind wings beat at different rates. Despite my distress, I felt strangely interested in observing this new phenomenon.

Without the forewing span that had balanced him like a tight-rope walker, Spot kept tipping to one side and then the other. Perceiving his agitation, I gently cupped him between my palms and he calmed down. He took the drink I offered. I took one, too. Scotch.

Spot learned to balance and steer using his hind wings. I placed him by a sunny window on a hibiscus plant for the support their broad leaves offered. He fed again in the evening and seemed healthy. But we had turned a corner. Although irreparably changed, he was still alive. I had taken over the natural progression of his life, hoping to improve his functional ability. With a deep sigh, I accepted the choice I'd made to try.

Failures often precede successes. Perhaps mine would lead to others' wins. I thought of the war survivors I'd seen, whose missing limbs were replaced with the best that science, medicine, and engineering could offer. I'd watched them with buddies at their sides when, in various ways, they ran in November's New York marathons.

February 5

Spot has lost the nob at the end of one of his antennae.

Spot would still have half his sense of smell. He was coping with his pruned wings, fluttering rapidly until he caught his balance enough to perch on my finger. His wings showed some pale spots where their scales had broken off. These were signs of more deterioration as he aged. But his legs were strong, and he still fed normally.

If he were in the wild, staying alive had meaning. He could still be food for other creatures. Invulnerable to his bitter taste, spiders would find him a tasty morsel. But not in the Virginia winter where it was icy cold.

I knew his days were numbered. Every day was a gift, an opportunity to learn, be happy, show kindness, and experience challenges with grace. Spot's life gave me pleasure when he was younger. Learning about and caring for him gave meaning to dreary pandemic isolation. As he aged, he showed me the courage and persistence it takes to accept life's changes, as I learned from a Buddhist monk during a vacation in Japan.

"There are three things you must learn.

First, everything changes.

Second, you are changing.

Third, you must accept the first two things."

Week 6

February 6

How long can he live like this?

My breath clouded before me when I returned home, hungry at seven from the late vaccination clinic. I found Spot, listing to one side and very still, on the floor. Preparing myself that he was probably dead, I bent to pick him up. He waved a leg.

Perhaps he'd gotten hungry and tried to flutter off the twig where I'd left him at noon. He began supping as soon as he got a taste of honey water. He wiped his proboscis before curling it away then perched upon his customary twig for the night.

The color of his wings was fading daily.

He still managed to stay alive.

I should attend to his increased risk of falling like with anyone aging.

February 8

Spot might have broken a leg.

After a day off, I was scheduled to give vaccines this morning. At six, when I had to leave, dawn hadn't yet arrived. The thermostat wasn't set to warm the house for another hour. It was so cold and early that Spot wouldn't eat before I had to go. When I got home at one, the house was warm but Spot was on the floor again. He was limp and weak. Crouching to get a closer look, I saw Spot barely waving an antenna as if to signal he was still alive. Feebly, he climbed onto my finger. His proboscis was extended with hunger.

His right front leg was folded tightly beneath him, possibly broken from his handicapped flight and crash landing. He drank for a long three minutes before curling his proboscis and tucking it away. I decided that in the future before going out for an extended period, I would pour honey water on a sponge and leave him on it.

February 9

Spot's left foreleg is broken. He can't extend the lower part to grasp and keeps it folded near his body. He drank a little but was barely able to curl his proboscis after finishing. More wing scales are flaking off. He lost the sensory tip of the other antenna, so he no longer has a sense of smell. He might not make it through the night.

February 10

Spot is listing to one side again, but still clings to life. Unable to drink, he barely flutters his wings. I decided to take him outdoors for his final rest.

I didn't have the courage to leave him there. I brought him back in and set him on the sponge with some honey water. Perhaps he'd muster the strength and desire to sup. I was in the bargaining stage of grief. I set him carefully with his legs propped to hold him upright. I hoped he was comfortable near the bright window he'd preferred when he was well.

Was his digestive system failing, too, like what happens sometimes when people are dying? When I checked back later and tried again, I still couldn't stimulate his interest in feeding. I placed him on his branch but soon found he had fallen off. I tried something I remembered he'd done earlier. He climbed onto the front of the bright, orange cable-knit sweater I wore and clung there. Taking the laundry out of the machine was a challenge. I accidentally knocked him off once. He fluttered wildly but was quick to climb back on. He calmed down. We stayed together that way until I took the sweater off carefully for bedtime.

ABBEY PACHTER

On my sweater

Week 7

February 13

Spot's strength is declining.

February 14

Valentine's Day

For the past few days, Spot's been very weak. He's only moved after being warmed near the fireplace. Then, as if for exercise, he'll rotate his head, rise a bit on his legs, and flutter furiously, briefly stretching the muscles to his very short, frayed wings.

Spot showed no initiative for nourishment. He'd lost both antenna tips and one foot. The other leg was damaged. I figured his anorexia was from decreased or nonexistent senses of smell and taste. Rather than just letting him starve—a value-laden expression—I tried to encourage him—a human trait like making up a lunch-sized plate with small amounts of attractively arranged, colorful food for a frail, old person with similar sensory deficits. I used a toothpick to unfurl his proboscis, then placed a honey water-moistened Q-tip under it. With that, he took a brief sip.

Had he been left in the wild, Spot might have drifted to a secluded spot and waited there until he died. He might linger, left alone, like some elderly people. With diminished hearing, vision, smell, and taste, they become withdrawn and isolated. Both have one foot here, the other awaiting death. Perhaps gradual withdrawal is nature's way of preparing us for death. Or, he might have met a less poetic end. He might have been eaten.

By laying about four hundred eggs, butterflies produce more caterpillars than can be nourished by the amount of milkweed naturally available. Attempts to raise monarch 'cats' by well-meaning amateur naturalists may result in the release of more butterflies than milkweed stands capable of supporting their offspring. Worse, unwittingly providing non-native milkweed inhibits migration and spreads deleterious infestations.

Caterpillars are primarily destined to become food for other animals. Carnivorous praying mantises (Mantodea), for example, hatch in the same season as caterpillars, which they eat. Insects attract birds, which feed their babies thousands of caterpillars and other insects. Nature provides protein essential to their healthy development. Nature also provides checks on overpopulation.

During the initial years of COVID-19, more people who didn't believe in the science of quarantine and later refused vaccines died than those who took such life-saving precautions. Sadly, many others died due to other health problems irrespective of their beliefs. As is natural with viruses, COVID-19 eventually evolved to become less lethal and more contagious. One might say we are co-evolving. Monarchs infected with *Ophryocystis elektroscirrha* (OE), which is not a virus but a parasite, live, reproduce, spread OE, and make more caterpillars and butterflies for it to infect. Infestation inhibits successful migration. These conditions promote the persistence of OE.[43] Perhaps the increasing levels of cardenolides in milkweed, which offer some protection against predation, will offer some resistance to infestation. Time and research are needed to answer this question.

Monarchs have other predators. Tachinid flies and braconid wasps parasitize them. Loss of contiguous habitat, climate change, and unintentionally harmful human activities take their toll on monarchs and promote other insects which are, interestingly, better pollinators. These are filling a biological niche needed with the decline of honeybees. We do well to appreciate every butterfly we are privileged to see.

February 15

The damaged foot has improved. He keeps it extended, which helps him stay upright.

I set Spot safely inside a cupped mask, where he was very still. Was it time? Would it be merciful to put him in the freezer to release him from his diminished life?

I still didn't do it. I couldn't easily explain why. My scientific training? Curiosity? Ruthlessness? Lack of feelings? Need for companionship? My emotional response to all the deaths due to COVID-19? Attitudes toward aging? Everything.

February 16

He's eating again.

It felt like a reprieve. Spot groped for nourishment for the first time in several days. Perhaps he'd needed time for his leg to heal or to adjust to his winglessness. And since he wasn't flying, he used fewer calories.

While resting quietly on a dry mask, he unfurled his proboscis and probed it looking for nourishment. Seeing this, I offered him honey water. He sucked some up even without much of a sense of taste or smell. He must have received other biological signals to search for nourishment.

He fluttered his wings a couple of times again and was able to move his body slightly. It appeared to take him great effort to dislodge his feet, which were hooked into the mask fabric. Sometimes I helped. His head was droopy the rest of the time as if holding it aloft required too much effort.

February 17

Joys and sorrows.

When I saw the small flakes that had fallen off Spot's wings that lay on the table beside him, I approached with a small, hand-held field microscope until I was barely an inch away. With a minute adjustment of the focus, myriad tiny, rainbow-hued scales suddenly appeared.

Knowing about wing scales was one thing. Seeing them magnified was entirely different. They were beautiful, iridescent—nothing less than another miracle of nature.

Several small spots of pink fluid stained the fabric under his face. The only other pink liquid I'd seen come out of a monarch was the waste expelled before one became a chrysalis. Digestive juice might indicate a blockage. Something wasn't working properly. Spot's abdominal section was very thin. He must have lost micrograms of weight. Perhaps this represented a loss of reproductive capacity. It was difficult for Spot to place his weakened legs to balance upright.

If I were in this condition, I imagine I would appreciate compassion. I set Spot onto a cashmere sweater I wore and moved his legs into a stable alignment. I set aside an hour to relax and read. He held on until I set him back in a cupped mask for the night.

Week 8

February 21

Fifty days. It is hard to comprehend that Spot is still alive. His condition seems stable.

I'd adjusted to his senescence. Perhaps he had, too. During the past few days, he drank morning and afternoon. Like a doting mother, or perhaps now more like an adult child caring for her mother, I encouraged him to eat and he'd warmed up near the fireplace. Sunny windows were rare on those short, mid-winter days.

Sometimes, Spot's wing remnants fluttered rapidly. The physics of it made sense: the same muscle power—though perhaps somewhat diminished with age—and less weight resistance meant a faster flutter. It happened at various times of the day—most often shortly after he ate.

I enjoyed watching my children play when they were young. Watching Spot's post-prandial excitement was similar. My older daughter was sensitive to concentrated sugar and red dye.

"How was the soda at your friend's home this afternoon?"

"It was great! How'd you know?" She answered my question while gleefully fluttering around our living room like a beautiful butterfly. My younger daughter, a social worker and triathlete, had other characteristics I saw in Spot—patience, courage, athletic strength, and persistent endurance. When I make a wish, it is always for my family's health, happiness, and longevity.

February 23

Before traveling, I thought again about the freezer, but couldn't do it.

I flew to Chicago for my sister's birthday. A friend took care of my dog. Spot couldn't be left alone for the ten days I planned to be away from home or he would starve to death. He traveled with me in a small, covered but vented bowl. I packed him into my carry-on computer bag. Once we were at flight elevation, I took him out for a look out the window. I chuckled while watching Spot look at the tops of clouds, a unique perspective for any butterfly.

Spot looking out the jet window

Friends who got to meet Spot laughed. He gave them a reprieve from their depression after months of COVID-19 isolation. He was fine after the flight and had an evening feeding once we were settled into my mother's 27th-floor condo, where he had a view of Chicago's skyline.

Spot looking at the Chicago skyline

February 25

To visit long-time friends, I'll drive Spot from Chicago to South Bend, past my childhood hometown. The dunes and oak forests along US-20 hold the roots of my love of nature.

Spot seemed to tolerate the different water and local honey I used at my friend's home to make him fresh food. He surprised me with more frequent and longer periods of unproductive fluttering, perhaps alerted by the change of scenery. Unable to fly, he walked with an uneven gait, steadied by his wing stumps that functioned like rudders on a sailboat. Perhaps the action lifted some weight off his tired legs. When he stopped, he adjusted his legs a bit for stability. On the drive back to Chicago, I pulled over and collected a few of the abundant, dry milkweed pods from the side of the road.

Visiting friends

Week 9

February 27

Spot is 8 x 7 x 2.86 = 160 human years. Much older than Moses but younger than Methuselah. Old age can last a long time. It is often hidden from sight and unfamiliar to us.

February 28

I hope he doesn't feel pain.

Spot's right front leg, broken a couple of weeks ago, didn't heal properly. He moved it very little in the past couple of days and couldn't grip with that foot. Perhaps his guarding the leg was a response to pain. I was glad I couldn't know how he experienced it.

I tried some physical therapy (PT) exercises. I gently, minutely stretched his diminutive leg muscles, extending them from their flexed position, then let them return to their bent position. Stretch and relax. Repeat five times. Twice daily. It didn't help right away, but human PT took time, too. Heaven knows it had taken time and lots of PT before I could walk without pain after a complete knee replacement several years ago.

Part 3

March

Curiosity is the beginning of knowledge.

—James Clear[44]

March 2

How much longer will he live?

Physical therapy was helping. Spot resumed anchoring himself to things again, using the tiny spurs beneath his right front foot. In the morning, it was still stiffly flexed but responded to gentle stretching by becoming more functional. However, Spot fluttered his deteriorating wings more furiously and more often. Perhaps he's losing some part of his butterfly brain that controlled flight.

How much time did we have left? Since he'd eclosed so late, I wondered if Spot would live until spring like his migrating cousins, which greatly outlive earlier hatchlings. The question reminded me of a discussion between an orthodox rabbi and my nursing students.

"Do Jewish people believe there's a heaven and hell?" They wanted to learn about elderly people they would be assigned to care for in a largely Jewish long-term care facility. He looked at them with great seriousness except for the twinkle in his eyes and gave an answer that brooked no argument.

"We'll all find out."

Would Spot live until spring? I'd have to wait and see.

Most monarchs rarely survive more than a few weeks. Others, identified by tags placed and collected by lepidopterists and citizen scientists, have shown that late fall-emerging butterflies could live up to nine months during migration.

Which kind was Spot? He had arrived late—after all the other monarch caterpillars had disappeared into the foliage of my small, suburban, front yard. The angle of the winter sun at noon had already dipped below fifty-seven degrees from the horizon—the angle that signals monarchs to start their annual, iconic migration[45] to Mexico.

Perhaps he was desperate to follow his biological urge to head southwest. The rest of the caterpillars that had disappeared might have hidden to wait out the winter.

March 3

Spot is fastidious.

When daylight warmed the room, Spot awoke and began fluttering wildly and persistently. I moved him to a brighter spot, but that didn't quiet him down. I moved him to the sunniest window ledge, to no avail. I finally noticed that rather than being tightly curled, his proboscis was stuck in the shape of a bigger circle than usual.

I tried to unstick it by touching it with honey water soaked into a Q-tip. He turned his head from side to side like a person indicating an emphatic "No!"

Uncertain of how to interpret his behavior, I started with what he'd taught me already. Trained well by now, I mixed a fresh batch of the honey water—as he preferred—then placed a clean, saturated Q-tip under Spot's right foot. Tasting it, he quieted down as if thinking, "That's better!"

But he still couldn't extend his proboscis to drink.

I tried another trick. I carefully probed the end of a toothpick inside its circle like I did when he first eclosed, two months ago.

Whether I'd unstuck it or he was irritated enough to overcome the intrusion, he immediately unfurled his proboscis and began to drink. Thirsty from the delay, he drank for a couple of minutes, which was longer than usual.

Then, he did something I never would have imagined.

With no tiny napkin in site, he wiped that hair-slim tube onto his abdomen several times, on both sides. Only then did he curl it away. I sighed with pleasure. He'd solved the problem of a sticky proboscis.

This was another miracle, another learning, another intimate connection I made with this small creature Nature entrusted to me. This was not simply an anthropomorphic analogy. He'd learned how to cope with the honey water he had to drink to stay alive in captivity. He didn't want his proboscis to get stuck again.

Spot aged with grace. Although at times I considered the freezer for a quick end, I couldn't. I wasn't afraid of his life ending in its own time. I'd helped my dad live the last weeks of his life and learned to accept the process. Helpful advice and medications from hospice nurses taught me how to ease the natural discomforts that arose near the end of life. I don't think my dad suffered through it. I hoped Spot didn't suffer either. Decay is part of life. Suffering arises from the human cerebral cortex, absent in butterflies. I continued to learn from Spot, especially during that late stage of his life.

Aging takes its own time. Excepting an untimely occurrence, it can be a slow process. Humans might have strokes, heart attacks, cancer, other fatal illnesses, or accidents. Insects could be eaten, infected, or frozen. Otherwise, life persists to its eventual, necessary conclusion. Each species has a defined lifespan, as far as we know today.

Each day with Spot required flexibility rather than a strict schedule, which suited my nature. Spot was hungry at various times of the day. Sometimes he rested longer than other times between meals.

March 5

I am pleased with the beautiful notecards my sister printed from the photographs of Spot.

One of Spot's photos showed him, newly eclosed, hanging upside down to dry his wings. The notecard revealed something I hadn't noticed. Although I hadn't seen it split open, I'd captured his earliest moments in that photo. His hind legs were still inside the translucent cuticle, the remains of his chrysalis. Noticing this for the first time thrilled me. The more one looks, the more one sees.

Another card showed Spot standing with his wings wide on the Japanese floral cloth napkin he loved. It looked like he was touching—wingtip to wingtip—the woven image of a crane as if he were a part of the landscape.

The notecard images brought sweet memories of his best days.

Week 10

March 7

64 days. Spot is travel-weary and shows no interest in eating.

Spot deteriorated further. He had central nervous system problems that caused him difficulty moving about. He fluttered wildly and moved erratically, backward, or in circles.

One afternoon, I found him on the floor. He'd fluttered off the bed. I gently moved him to the center of the bed and surrounded him with supportive, soft washcloths, so he wouldn't harm himself. I'd need to safely corral him from now on.

We flew home.

More scales flaked off his wings, decreasing their size even further so that they no longer overlapped. His forewings moved very little. Whatever neurons coordinated them had failed. His hind wings didn't move at all. Spot made a little frantic fluttering in the morning. I offered him food. He wouldn't eat.

I'd learned something about caring for others who neared the end of life. People often refused food, which distressed their loved ones.

"If only she'd eat something, she'd get stronger," my father said about my grandmother during what became her final days.

As gently as I could, I told my parents that Grandma wouldn't be getting any stronger.

"When people are nearing their passing," I said, "their bodily systems, including digestion, start to fail. They stop eating because it is more uncomfortable to eat. Grandma's OK." Sometimes she drank small sips of water. That continued for many days.

After this discussion, my parents turned a corner in accepting Grandma's decline.

I turned a corner accepting that Spot had little time left although I didn't know how long he might linger.

"People don't come with expiration dates," someone wise told me dozens of years ago. Neither did Spot.

On our weekly Zoom call, my grandkids wanted to see Spot—they always did. I held him gently so they could see his fragile condition. I told them Spot had grown very old and probably would not live much longer. The circle of life was turning.

March 10

Spot refuses a second meal after eating only once, early this morning.

March 12

Spot has lived seventy days. He hasn't eaten since early yesterday. Sometimes he frantically flutters the remains of his wings. He can't move forward in any coordinated way. I think he has dementia in the actual meaning of the word: de-mental. His brain is no longer directing his movements.

Intuitively, I emptied one of the sock-filled shoe boxes from a dresser drawer and carried it outdoors. Searching for something suitable to keep him safe, I gathered a bunch of leaves that had fallen leaves under the pear tree. The leaves were sun-warmed, soft, and dry in shades of warm oranges and browns. They closely matched Spot's now-muted colors. I patted down the leaves into the shoebox and gently placed Spot on them. He immediately calmed down, perhaps knowing he was well camouflaged from predators. I hoped he felt cradled and supported so he could die in peace.

He aroused to take the longest drink ever. Perhaps the long drink, protracted to almost five minutes, was further evidence of his dementia. Then he lay still.

Perhaps that would be his last effort.

I reflected on the days when he was young, and a drink would give him an energy surge. What a joy it had been to watch him exuberantly fly up to the ceiling fan, skylight, and windows. In contrast, this sup led to a very long nap.

Week 11

March 14
Morning

Sunday—no vaccine clinics today. Spot barely moves and rarely eats.

I checked on him first thing in the morning—as I have for eleven weeks. Spot was quiet in the leafy habitat where I'd put him two days ago. The difference was that Spot's proboscis was unfurled as if he were hungry. When I tried feeding him, he tried, too, but his movements were feeble. His wings fluttered just a little bit. He briefly probed the honey water. He was too weak to do more. He seemed calm for the first time in a week. I didn't want to disturb that calm, so I didn't touch him.

I was certain he was experiencing his final hours. He seemed comfortable and quiet in the open shoebox. I set it in the warm sunshine among the other leafy, winter detritus in my backyard garden.

It was a beautiful day. I felt at peace and was glad that Spot seemed in no distress. Everything was as it should be. He'd lived much longer than anyone might have expected. His life was taking its natural course. So was mine. I began spring cleaning.

Midday

I walk my dog on the Chesapeake Bay beach a couple of blocks from home. Although it's still chilly, spring-break sunbathers lie on colorful towels to add warm tones to their pale skin.

While waiting for a cup of soup to heat, I stepped out into the garden to check on Spot. I felt a different kind of stillness. He lay motionless among the leaves. I touched a bit of wing to make sure.

He had died.

I felt a mixture of sadness and relief. The disturbing memories of his disorientation and panicked frenzy of recent days confirmed my suspicion that his time was up. Those images passed quickly. A soft calm enveloped me. I'd had anticipatory grief—like when parents, recently bereaved, said they knew something was amiss. I'd been preparing for this day for weeks.

"To love what death can touch...."[46]

I took a deep breath. Then another. I felt one last responsibility to Spot.

I cleared a deep patch and set the leafy contents of the shoebox, along with Spot's remains, into the rest of the leaf mulch in my garden. It was already becoming nutrient-rich, fresh soil. Better than dust to dust. I covered him with a few of the nearby leaves and started indoors. When I took one last look back to where I'd set him, I couldn't distinguish his final resting place from its orangey-brown and golden surroundings. That felt right and good.

RIP

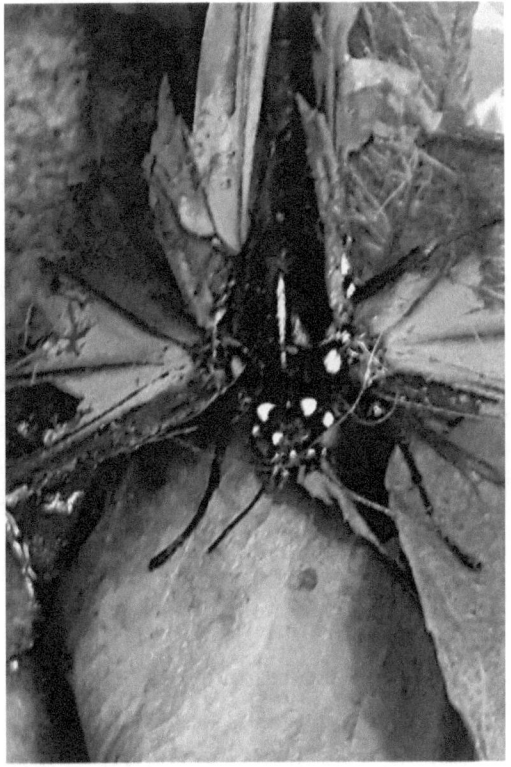

Evening

    I put a notice of his passing and the last photo on my Facebook page.

I spent the rest of the day in quiet contemplation of life in all its forms—the web of nature. Looking back at his photos, I smiled and recalled the joys of Spot's young life. He'd given me lessons in caring right away and lessons in patience as he aged. I'd learned how to feed him and respond to his changing needs. I'd been surprised to discover so many evolutionary similarities between insects and humans. He was a companion who rested while clinging to my bright, orange sweater. He was there for me to greet when I returned home from those intense hours at the convention center giving hundreds of COVID-19 vaccinations. It's one kind of love.

March 15

It comforts me to read the notes friends sent.

I awoke a few hours before my shift would start at the convention center. Before getting up, I read dozens of condolences posted to me on Facebook. Many friends had met Spot. More had seen photos and heard about our interactions during the past several months, and had grown attached.

"Thank you for sharing Spot's journey with us."

"I'm sorry for your loss."

"I never knew so much about a butterfly's life."

Several mentioned having a new appreciation for and connection to a tiny part of our amazing, interconnected, and beautiful, natural world. Their comments made me glad to have shared my experiences. I'd learned to slow down, observe, learn, keep studying, and keep writing. Spot enriched my life and had meaning to others.

March 17

Grace surrounds me in its soft, cottony embrace. I feel honored to have shared his life.

I hoped that with so many people getting vaccinated, our human activities could soon resume some semblance of normalcy. After completing the week's clinic schedule, I caught up on email. A sweet friend sent a link to the Monarch Watch website. Monarchs were beginning to return north from their winter roosts in Mexico. Maybe I'd go there to experience it first-hand.

Some had been sighted in Texas. Their cycle of life was beginning again.

In Memoriam

Appendix A

The Butterfly Life Cycle

Monarch butterflies (**Danaus plexippus** begin as tiny eggs laid by female, adult butterflies. Each species of butterfly egg has a unique shape. Monarch butterflies' pinhead-sized eggs have longitudinal ridges from pointed tip to round base. Although some butterflies will lay eggs on a variety of **host plants,** most will only lay their eggs on one species of plant. Monarch butterflies only lay eggs on varieties of one species: milkweed (**Asclepius**). About three days after being laid, eggs hatch as **larvae** (larva, singular). Larvae, commonly called **caterpillars**, eat only milkweed leaves. They grow and then shed their outgrown skin in five stages over two weeks. Each stage lasts only a few days and is called an **instar**. All butterflies do this, but all have a unique appearance to each instar.

During their fourth instar, monarch caterpillars attach to a twig. They use their **cremaster**, which is an appendage covered with tiny hooks—like Velcro. They expel a tiny pad of sticky exudate onto the selected twig and twist their cremaster into it until it is firmly attached.

After a day's rest, caterpillars lose their legs and split their skin for a final, fifth time. This reveals a brief grub stage called a **pupa** (plural pupae), which quickly dries to have a hard shell. Then, it's called a **chrysalis**. This name dates to 1600 from the Latin *chrysalis*, derived from the Greek *khrysallis*, from *khrysos* "gold," and suggested a sheath. **Chrysalides** (n. plural) and **chrysalid** (adj.). are derivatives from 19th century French.

Monarch **chrysalides** are Kelly green. They have a collar of tiny, gold spots with several more scattered below. Chrysalides are distinguished from **cocoons**. By naming convention, moth pupae become cocoons. Butterfly pupae become chrysalides.

The pupa inside the hard case dissolves into a genetic broth that is reconstructed through a still-mysterious process called **metamorphosis**. After about two weeks, primarily dependent on ambient temperature, adult butterflies **eclose** (emerge, hatch) from the chrysalis. They crawl out of all that's left of the chrysalis--its clear **cuticle**. The still-compressed "newborns" must hang with their wings down. Gravity distributes a fluid, plasma-like **hemolymph** from their body to expand the wings to their adult shape. Any handling or disturbance that interrupts this process can render them useless. It takes about two hours until they are fully dry, hardened, and able to sustain flight. Then, butterflies take off to find food from **nectaring** (nectar-producing) flowers.

Males will quickly find and mate with females, which immediately start laying eggs. In two to three days, the eggs hatch to start a new cycle of life.

Females **eclose** carrying about 400 eggs, which they deposit in the next couple of weeks if they aren't gobbled up by predators. Human females carry about 400 ova (eggs) at birth. *The same number.*

Appendix B

Nectar Alternatives

Honey is closest in nutrients to nectar from flowering plants, the natural food of adult butterflies. After Spot rejected Gatorade, I investigated its nutrients. Gatorade was not designed for basic nutrition but to replace electrolytes lost in sweat during human exercise. It lacks the diversity of vitamins and minerals present in honey.

I compared honey to sugar, maple syrup, and synthetic honey and found no alternatives with as close a profile to nectar as honey. Both honey and nectar are primarily composed of sucrose, a polysaccharide bonded from fructose and glucose. Nicolson reports that "the energy from its sugars, while crucial, is not the only benefit" of nectar. Its other components, while highly variable across species and environmental conditions, most likely contribute significantly to (pollinator and butterfly) nutrition. These include water, minerals, and amino acids for protein and taste, fatty acids for fuel, Vitamin C that retards spoilage, and secondary metabolites, which inhibit microbial and yeast growth that otherwise consume the sugars.[47] She concludes that availability and wide floral diversity should provide optimal nutrition.

I found a variety of dilutions posted online from 1:4 to 1:9. None had research support. The 1:6 dilution of honey to water that I used provided enough calories and nutrients for Spot. It was thin enough to permit capillary action and sucking and contained enough honey to provide sufficient nutrition for his sustenance. Perhaps 1:4 or 1:5 would have prevented his "fainting" episodes but that concentration might be too viscous for him to suck. This would make an interesting experiment for monarchs raised for scientific experiments. Otherwise, captive rearing is discouraged.

Flowering plants evolved much later than sugar maples. Butterflies coevolved with flowering plants, suggesting to me that nutrients in nectar should be more specific for monarchs.

Appendix C

Taxonomy of Monarch Butterflies

Danaus plexippus

Did King Phillip come over for great spaghetti? This is one of the mnemonics used to remember the first letters of a naming system biologists created to categorize living things. The categories flow from general to specific evolutionary and genetic characteristics. Every form of life that's been identified falls under a domain (most general characteristics), kingdom, phylum, class, order, family, genus, and species (most specific). Look up any living thing to find a table showing its taxonomy, like the table included below for monarch butterflies.

All butterflies belong to the domain, Eukaryota. Humans are eukaryotes, too, because they are all living things made of at least one walled-off cell. Each eukaryote cell has a distinct nucleus, which contains the deoxyribonucleic acid (DNA) required for it to reproduce. Cell nuclei also contain organelles, including mitochondria, which provide energy. Viruses are excluded because they don't have DNA and use that of organisms they infect to reproduce. We belong to the same animal kingdom, because we all are multicellular (unlike amoebas) and we have muscles. Beyond that, we differ.

About 565 million years ago, there was a great divergence in the animal kingdom into two branches called phyla (phylum is singular). These are Chordata (vertebrates) and Arthropoda (invertebrates). There are about 60,000 species of vertebrates including mammals like us, birds, fish, reptiles, and amphibians. All have interior (endo-) skeletons including the vertebrae that protect the spinal cord, the nerve system that lies within. Due to their bony (or, as in sharks, cartilaginous), internal support structures, vertebrates can grow larger than invertebrates. Butterflies are in the phylum Arthropoda, which includes invertebrates.

In contrast to Chordata, there are about two million species of invertebrates. Instead of interior skeletons and vertebrae, they have exterior (exo-) skeletons which, unlike bone, are made of chitin, a protein somewhat like the keratin of our hair and fingernails. Nearly half of them are insects.

The class, Insecta, includes arthropods with bodies that are segmented and pinched between the head, thorax, and abdomen. They have compound eyes and a pair of antennae. Monarch butterflies are in the class Insecta, and more specifically, the order Lepidoptera. This order includes butterflies and moths, which undergo complete metamorphosis. Lepidoptera includes the family, Nymphalidae, which includes only brushy-footed butterflies. To further distinguish them, monarchs belong to the subfamily, Danainae, which have regressed forelegs, and the genus, Danaus, which are described as tigers for their orange and black markings. Their final distinction gives them the species name, Danaus plexippus. Their name was derived from Greek, meaning "sleepy transformation." The most familiar butterfly in North America, members of this species require milkweed plants (Asclepias) upon which to lay their eggs, because that is the primary plant species monarch caterpillars will eat.[48]

Adults use something like this on long car drives with children. The first player asks, "I spy, with my little eye, something...." To discover the answer, someone else asks, "Is it an animal? Does it have wings?"

APPENDIX C

Monarch Butterfly

Danaus plexippus

Level	Name
Domain	Eukaryota
Kingdom	Animalia
Phylum	Arthropoda
Subphylum	Hexapoda
Class	Insecta
Order	Lepidoptera
Superfamily	Papilionoidea
Family	Nymphalidae
Subfamily	Danainae
Genus	Danaus
Species	Danaus plexippus

Appendix D

About Milkweed

Two years after Spot died, I learned about an important error I made, from which I hope others will learn. Monarchs should only feed on locally native varieties of milkweed. In Virginia, these include Aesclepias syriaca (common milkweed), A. incarnata (swamp milkweed), A. tuberosa (butterfly weed), and A. verticillata (whorled milkweed). In contrast, A. curassavica (tropical milkweed) yields monarchs with structural changes that inhibit or prevent their iconic migration from the USA to Mexico. In warmer climates where it doesn't die back, the persistence of A. curassavica has been a significant driver of the dramatic increase in parasitic infection with OE. Spot fed on tropical milkweed, which I grew from gifted seeds before knowing it was problematic.[49] Perhaps that is what prevented him from migrating. Mea culpa.

Intervention in monarchs' natural life is not necessary. Contrary to what is nearly legendary, monarchs are abundant, according to many scientific studies. Domestic rearing or cultivating of monarchs by well-meaning amateurs is counterproductive.

Pankau (2023)[50] called it an "ecological trap," He affirmed that consumption of tropical milkweed in North America is inhibiting the long-admired, iconic migration of Monarch butterflies to Mexico. Although the overwintering population there is declining, it is not clear what mix of changes is causing this. Loss of habitat and increasing temperatures are contributing factors.

More information about milkweed is readily available.[51]

Acknowledgments

When we're young, our experiences seem discreet and uniquely of the present moment. Only as we age can we see events stitched together to make a whole cloth. I gained a deep appreciation of this while experiencing life with Spot. The threads from which the fabric of my life is woven came first from my parents, Louis and "Dolly" Pachter. They raised me in Gary, Indiana at Miller Beach. I faced Lake Michigan every day of my childhood. Monarchs and all manner of arthropods and flora were summers' accompanists.

My next-door neighbor, Margie Kohn, was an original naturalist with white hair and a no-makeup, beach-appropriate "pedal pushers" approach to most days. Her love of the northwest Indiana dunes predated all Save-The-Dunes efforts that resulted in their recent designation as a national park. Her husband, Stan helped me catch my first fish, a coho salmon, which I grilled for our families' dinner after I thanked its departing spirit for nourishing us.

Across the street, Mr. H. tolerated me as a six-year-old who quietly sat nearby while he tended his garden. He taught me how to dig up iris in the fall, prune their corms and replant them to invigorate showy, spring flowers. I was deeply moved when he gave me one. Since I planted and tended it well, he gave me orange daylilies when I was seven, the first native plants for my nascent garden. My father had created it alongside our driveway and enhanced it with a peach tree at one end and a pear at the other. Mr. H. added a few yellow primroses, which annually spread ten-fold. I transplanted several to my college garden and their offspring to every home I've had since then.

Charles Eilber led a nature appreciation course in Michigan at a small museum of natural history on the boy's campus of the Interlochen

Arts Camp (formerly the National Music Camp), now part of the Interlochen Center for the Arts. I was an eight-year-old camper and won my first-ever award for a pressed and labeled leaf collection. I still have it—a small display box of several types of rocks sliced and polished to reveal their inner beauty and labeled like my leaf collection that earned it. Six years later, he remembered me when I arrived at Interlochen for my junior and senior years of high school. He continued to be a kind and gentle role model as director of the Academy.

At about the same time as that camp experience, my parents learned that I had a significant hearing deficit. They selected the Kuhn Clinic in nearby Hammond, Indiana to test my hearing. While waiting with my mother, I inspected the large display cases that covered its walls and contained objects removed from kids' ears, noses, and throats. I can see those objects in my mind's eye as if it were yesterday, not sixty-some years ago. Open safety pins, beans, buttons, and baby rattles—I was fascinated that those large items could fit into such small spaces (I never did that). The physician showed us a model of the ear, pointing to where my damaged inner ear's acoustic nerve was located. He told my mother that nothing could be done to improve my hearing. It was my first view of the inside of the human body. Fortunately, tests showed my other ear had normal hearing and I was mainstreamed at school. It was twenty years until I tried a hearing aid, which changed my hearing from mono to (less than perfect but) stereo—despite many specialist's reports to the contrary. My hearing was markedly improved. When the director of my midwifery program insisted that I wear it all the time, I grew to depend on it and learned that nurses are sometimes more effective than doctors.

Recognizing my innate curiosity, Dr. Bill Deutsch, our family ophthalmologist, showed me a model of the inside of an eye at a normal checkup when I was about ten. It makes sense to me now that I noticed it on his desk—probably my eye level then—having been shown the ear model earlier at Kuhn.

Nine years later, I decided on a career in nursing. It was not a childhood fantasy but an artist's dream. Bored in college as an art major after many years of wonderful art lessons at the Junior School of the Art Institute of Chicago and by the advanced high-school art education I received at Interlochen, I was captivated by the life-sized model skeleton from which my college roommate studied. She was studying human anatomy from real cadavers, which was highly desirable but not available to art students. That did it. At Thanksgiving break, I told my parents I wanted to switch majors, from art to nursing. They were delighted that I would create a meaningful career suited to my interests that would provide me with some level of financial security. The addition of science to my upbringing in nature and my love of art was set.

I went on to use other models as instructional aids. In my late twenties, preparing to teach an embryology lesson to student nurses as an instructor at the University of Michigan (UM) School of Nursing, I found Dr. Alfonse Burdi at the UM School of Medicine. He generously donated three fetuses preserved with formaldehyde in glass jars for me to use as teaching aids. They had been spontaneously miscarried at significantly different developmental stages. He studied them to research the embryologic development of the jaw. Dr. Burdi stipulated that they be treated respectfully and returned, when no longer needed, for use by others or proper burial.

In my later thirties, I had the support of Marge Jackson, then Director of Nursing at UM, to move to the University of Philadelphia for doctoral study. Once there, I learned to write, thanks to Dr. Carol Germain. I learned, applied, and taught others the importance of information literacy. My chair, Dr. Florence Downs, encouraged my creativity, laughed a lot, and paid my tuition and stipend from Penn's deep pockets. Dr. Downs was my north star through what my mother called "the busy years." For me, they were filled to overflowing with science, research, writing, art, midwifery-with its myriad on-call hours, and raising my wonderful daughters.

Approaching retirement after a forty-year career, I thought about what I'd want to do with my free time. I learned about Master Naturalists from Barb Johnson, a friend and member of the Tidewater chapter. I signed up and loved the coursework and field trips that led to certification. It was right up my alley, an introduction to all aspects of our local environment. I logged volunteer hours at Norfolk's Botanical Gardens Butterfly House, founded and run by Ms. Lauren Tafoya.

A few years later, while journaling daily during the first six months of COVID-19, I discovered that I enjoyed writing as a daily practice. As the saying goes, when the student is ready, the teacher appears. I enrolled in an online memoir course taught by author and performance artist, Alison Wearing. I am indebted to her for learning more about the art and craft of writing and for the opportunity to meet many writers at a retreat she led in Italy the following year. Many of us continue to meet regularly over our common interests in writing.

In a private conversation, Dr. Peter B. Schultz, Emeritus Professor of Entomology from Virginia Tech suggested I read Chapman's "*The Insects*" (op. cit., 2013). It has been a valuable resource.

To all of those mentioned above, both living and deceased, I am grateful and give many thanks. They all continue to inspire me as I seek to grow as an author.

Special thanks to Linda Cobb, from the Virginia Beach Writers. She became a dear friend and thoughtful editor. Many thanks to the Virginia Beach Writers, Hampton Roads Writers, and Virginia Writers Club for their friendship, support, listening, and editorial assistance.

During the last phase of editing this book, I encountered many Facebook sites about monarch butterflies. Dr. Andy Davis, moderator of "The Thoughtful Monarch" Facebook group, consistently offered discussion and links to peer-reviewed, published, scientific research on this subject. I am indebted to Dr. Davis for providing the resources that have furthered my understanding of monarchs, native milkweed, and the dangers of misinformation. His work reassured me that monarchs are not becoming extinct. Concerns persist that their iconic migration is diminishing. Any errors in this book are solely mine.

Many thanks to Madeline for her support and creative assistance in selecting the cover design and for producing my headshot. The joy I feel shows- it comes from gazing at her during our shoot.

Many thanks to my project team at Paper Raven Books.

Postscript

My decision to bring a monarch caterpillar indoors started as an instinctive response. I had no idea what I'd do with him. But the impulse grew into something else—a desire to learn.

Winter has always been a more internal time for me, but the social restrictions I followed during COVID-19, when Spot arrived, created time for new levels of contemplation. Learning about this tiny creature so that I could care for him occupied me intellectually and nourished me emotionally. It took my mind off the fear and worries about the ongoing pandemic.

As I got more involved with his care, I gained perspective and unconsciously developed the emotional distance typical of nurses and doctors who, while caring for patients, do their best to be personable while maintaining healthy boundaries. This emotional space helps us avoid running out of the energy we marshal to assess problems, make educated decisions, develop care plans, and implement and revise them as health issues evolve. We know life is finite. We know our patients are only in our lives for a short while. Our job is to care for them with compassion during that brief time.

For many acute care medical and nursing providers, that perspective was overwhelmed by the intensity of caring for previously healthy COVID-19 patients, thousands of whom died while in their hands. One way many retired physicians and nurses like me could help was to give vaccines. We felt a need to pitch in, and a duty to the public. When vaccines first became available in January 2021, the Virginia Beach Medical Reserves Corps (MRC) began implementing a plan to massively scale up the administration of the vaccine. In an important part of the process, the Governor allowed the Virginia Board of

Nursing to temporarily re-license recently retired Registered Nurses specifically to engage them in vaccine administration. After applying to the MRC and being accepted into this re-licensure, I took several online courses required to join in the effort.

My participation in the vaccination effort coincided with Spot's later life. I felt like every parent of a puppy—at some point, he could be left alone safely for a while. Like almost everyone who acknowledged the pandemic, I had "cave syndrome"—fear of leaving home and contracting COVID-19, which was still deadly. It felt good to get out while wearing a mask and observing the suggested six feet of social distance as much as possible, to get past cave syndrome, and to help.

Along with dozens of other RNs locally and probably thousands nationally, I was awarded a Presidential Award for Volunteer Service (bronze) by President Joe Biden for my service.

Spot's death coincided with new demands on my time. March became April and I needed to care for my mother in Florida. I felt torn about leaving the vaccination clinics because of my feelings of responsibility to my community. Then I remembered my core value: family comes first.

Image credits

The author took all the photos unless otherwise attributed.

Human lungs: https://en.wikipedia.org/wiki/Lung/

Cartoon: "That's an old photo" by Tim Whyatt, used with permission.

Spiritual Nourishment: won OST 1st prize photograph, 2022: Remote Shabbat service. The image includes Rabbi Roz Mandelberg, used with permission.

"The Gates" installation in Central Park: "The Gates" Installation by Christo and Jeanne-Claude. Reproduction of a watercolor of Gapstow Bridge. Artist unknown.

Endnotes

Author's note: For these endnotes, I followed Chicago Style formatting unless the publication required otherwise. Scientific articles often follow APA style.

[1] *Call the Midwife.* 2013. Season 2, Episode 2, "Bradley's Basement." Directed by Roger Goldby. Written by Harriet Warner. First aired on January 27, 2013, on Netflix.

[2] Reginald F. Chapman, *The Insects: Structure and Function*, 5th Ed. (UK: Cambridge University Press, 2013), Chapter 17 "Gaseous exchange," 514. doi: 978-0-521-11389-2.

[3] Kimmerer, Robin W, *Braiding Sweetgrass.* (USA: Milkweed Editions, 2013).

[4] Retrieved from https://1.wikipedia.org/wiki/Animacy

[5] Retrieved from https://en.wikipedia.org/wiki/Lost_in_Space 2018. Directed by Irwin Allen and John Williams, aired on Netflix.

[6] Kissinger, Schmidt & Huttenlocher, *The Age of A.I. and our human future.* 1st Ed. (USA: Little Brown & Company, 2021.

[7] Retrieved from https://www.reconstructingjudaism.org/dvar-torah/choose-life/#

[8] David Quammen, *The Song of the Dodo: Island biogeography in an age of extinction*, NY: Scribner (1996). ISBN 0-684-80083-7.

[9] A. Majewska, A. Davis, S. Altizer, and J. de Roode, "Parasite dynamics in North American monarchs predicted by host density and seasonal migratory culling." *Journal of Animal Ecology*, 91(4), 2022. 780-793; S. Altizer and K. Oberhauser. "Effects of the protozoan parasite ophryocystis elektroscirrha on the fitness of monarch butterflies (Danaus plexippus)." *J Invertebr Pathol.* July 74 (1) (1999): 76-88. doi: 10.1006/jipa.1999.4853. PMID: 10388550; Retrieved from https://www.monarchparasites.org/oe#:text=Ophryocystis%20elektroscirrha%20(OE)%20is%20a,the%20same%20characteristics%20as%20animals.

[10] Retrieved from https://www.honeybeesuite.com/bee-secrets-what-happens-when-bees-make-honey/#h-five-major-steps-for-making-honey-from-nectar

[11] Retrieved from https://end.m.wikipedia.org/wiki/George_Boole.

[12] Aaron M. Cypess, "Reassessing human adipose tissue." *N Engl J Med*, 386 (2022):768-779. doi: 10.1056/NEJMra2032804; Deborah Hoshizaki, Allen Gibbs, Nichole Bond, "Fat Body," in *The Insects Structure and Function*, ed. S. J. Simpson and A. E. Douglas (Cambridge University Press: 2013), 136-7.

[13] Retrieved from https://en.wikipedia.org/wiki/Franz_Naegele#: ~:text=Franz%20 Karl%20Naegele%20(7%20December,was%20also%20a%20noted%20obste- trician.&text=He%20earned%20his%20medical%20degree,a%20medical%20 practice%20in%20Barmen.

[14] Retrieved from https://en.wikipedia.org/wiki/Carl_Friedrich_Gauss

[15] Retrieved from https://www.investopedia.com/terms/b/bell-curve.asp#:~:tex- t=A%20bell%20curve%20is%20a%20graph%20depicting%20the%20normal%20 distribution,relative%20width%20around%20the%20mean.

[16] Retrieved from https://www.britannica.com/biography/Antonie-van-Leeuwenhoek

[17] Harald Krenn, "Functional morphology and movements of the proboscis of Lepidoptera (Insecta)," *Zoomorphology* 110, (1990): 105–114. https://doi.org/10.1007/ BF01632816

[18] Retrieved from https://www.thoughtco.com/how-do-insects-breathe-1968478

[19] Douglas Blackiston, Adriana Briscoe, Martha Weiss, "Color vision and learning in the monarch butterfly, Danaus Plexippus (Nymphalidae)," *J Exp Biol.*, 214, Feb. l, 2011 (Pt 3): 509-20. doi: 10.1242/jeb.048728. PMID: 21228210. https://pubmed. ncbi.nlm.nih.gov/21228210/

[20] Paramahansa Yogananda, *Autobiography of a Yogi*. (Los Angeles: Self Realization Fellowship, 1998).

[21] Journal of the Virginia Writers Club The 2021 Golden & Teen Nib Fall Edition Vol- ume 2, no. 1. Page 33. https://www.amazon.com/Virginia-Writers-Club-Golden-Teen/ dp/B09NRZM27X/ref=mp_s_a_1_2

[22] Retrieved from https://www.akc.org/expert-advice/health/ how-to-calculate-dog-years-to-human-years/

[23] Retrieved from https://doi.org/10.1098/rsif.2011.0392. Daria Monaenkova, Matthew S. Lehnert, Taras Andrukh et al. (2011). "Butterfly proboscis: combining a drinking straw with a nanosponge facilitated diversification of feeding habits." (UK: Journal of the Royal Society Interface 17 August 2011).

[24] Hoshizaki, Gibbs & Bond in Chapman (2013), The Insects, 136-137.

[25] Kathleen M. Lucas, James F. C. Windmill, Daniel Robert, Jayne E. Yack. "Audi- tory mechanics and sensitivity in the tropical butterfly *Morpho peleides* (Papilionoidea, Nymphalidae)." *Journal of Experimental Biology* (2009) 212 (21): 3533–3541. https:// doi.org/10.1242/jeb.032425.

[26] Hebrew abbreviation for the expression, "May his memory be for a blessing."

27 Retrieved from http://en.wikipedia.org/wiki/1966_Florida_Gators_football_team#

28 Crossley, M. S., Meehan, T. D., Moran, M. D., Glassberg, J., Snyder, W. E., & Davis, A. K., "Opposing global change drivers counterbalance trends in breeding North American monarch butterflies." *Global Change Biology*, 28, (2022), 4726–4735. https://doi.org/10.1111/gcb.16282

29 Cowie, R.H., Bouchet, P. and Fontaine, B., "The Sixth Mass Extinction: fact, fiction or speculation?" *Biol Rev*, 97, 2022: 640-663. https://doi.org/10.1111/brv.12816

30 C. Finn, F. Grattarola, and D. Pincheira-Donoso. "More losers than winners: investigating Anthropocene defaunation through the diversity of population trends." *Biol Rev.* 2023. https://doi.org/10.1111/brv.12974

31 M.R. Kendrick, J.W. McCord, "Overwintering and breeding patterns of monarch butterflies (*Danaus plexippus*) in coastal plain habitats of the southeastern USA" *Sci Rep* 13, 10438, 2023, 53. https://doi.org/10.1038/s41598-023-37225-7.

32 Quammen, *Song of the Dodo* (1996).

33 Anurag Agrawal. *Monarchs and Milkweed: A Migrating Butterfly, a Poisonous Plant, and Their Remarkable Story of Coevolution.* (NJ: Princeton University Press, 2017)

34 Retrieved from http://news.bbc.co.uk/earth/hi/earth_news/newsid_8481000/8481380.stm?fbclid=IwAR34l5bsldtDWzwcyzMC4Xx-kUUDd-JGIetNX74MLVXS2k6Qrcwj1ONqlhdM/

35 Retrieved from https://en.wikipedia.org/wiki/Thomas_Fairchild_(gardener)#:~:text=Thomas%20Fairchild%20(%3F,still%20denied%20by%20most%20botanists.

36 Retrieved from https://www.linnean.org/learning/who-was-linnaeus/young-linnaeus

37 Retrieved from https://www.britannica.com/biography/Gregor-Mendel

38 Retrieved from https://whitney.org/collection/works/22792_link/

39 Douglas Blackiston, "Color vision," 2011.

40 Retrieved from http://www.onezoom.org/.

41 Retrieved from https://www.monarchscience.org/single-post/new-study-published-despite-winter-colony-declines-monarchs-are-thriving-in-north-america-really?fbclid=IwAR3buTYg5r8u872WnDr5KixGAZjNWALmz9jZNkCNCchcocFqRX-rz-L_E2ZQ/

42 Retrieved from https://mathsciencehistory.com/wpcontent/uploads/2020/03/132_kap6_lorenz_artikel_the_butterfly_effect.pdf

43 Retrieved on 8/25/2023: https://fb.watch/mEh66IYmKY/?mibextid=Nif5oz

44 James Clear, *Atomic habits: an easy and proven way to build good habits and break bad ones.* Penguin: Avery, 2018.

[45]   Retrieved from https://news.yahoo.com/spends-days-counting-monarch-but-terflies-140024118.html

[46]   Attributed to Chaim Stern (1930-2001).

[47]   Nicolson SW., Sweet solutions: nectar chemistry and quality. Philos Trans R Soc Lond B Biol Sci. 2022 Jun 20;377(1853):20210163. doi: 10.1098/rstb.2021.0163. Epub 2022 May 2. PMID: 35491604; PMCID: PMC9058545. http://doi.org/10.1098/rstb.2021.0163

[48]   Retrieved from https://www.itis.gov/servlet/SingleRpt/SingleRpt?search_topic=TSN&search_value=117273#null.

[49]   Retrieved from Facebook group, The Thoughtful Monarch, hosted by Dr. Andy Davis, with references: https://www.monarchscience.org/single-post/a-complete-sum-mary-of-the-previous-and-new-science-on-tropical-milkweed-and-monarchs?fbclid=I-wAR2OTD0PV5z6t81WG4ADIv4bGwQ0cCoZqBHNyEQ3fT53ZILsFK0l---LBh0

[50]   Ryan Pankau, *Tropical milkweed could threaten monarchs*, The Garden Scoop, College of Agricultural, Consumer & Environmental Science, University of Illinois Extension, Urbana-Champaign, Illinois, July 22, 2023.

[51]   Retrieved from https://www.birdsandblooms.com/gardening/attracting-but-terflies/milkweed-guide/?fbclid=IwAR3bFvTQZwCycqDlplOkrkJxyMQYOe-KO7Fon7Dm5k15Au95fSDxfaUOYxF4

Author's Biography

Dr. Pachter was born in Chicago in 1952 and spent her early childhood enjoying nature in the Indiana Dunes east of Gary, Indiana at the southern tip of Lake Michigan. She graduated as a fine arts major from the Interlochen Academy for the Arts and earned degrees in professional nursing from the University of Michigan (BSN and MSN) and the University of Pennsylvania (PhD). Her career included clinical practice in nursing and midwifery, project management in healthcare informatics, and faculty and administrative appointments at several universities. She has been an active participant on many Boards of Directors and participates in several writing communities.

In addition to her dissertation, she has published several articles in healthcare-related journals. Her writing is most often based on personal experiences, inspired by nature, and informed by science. Dr. Pachter won first prize for non-fiction in 2021 from the Virginia Writers Club.

The author enjoys travel, and most recently visited Alaska, Italy, Egypt, and Canada. She is a mother of two, *Gramma* to four, and lives in Virginia Beach.

Index